Communications in Asteroseismology

Volume 154
June, 2008

Proceedings of the
Delaware Asteroseismic Research Center and
Whole Earth Telescope Workshop
Mount Cuba, Delaware, Aug. 1-3 2007

Edited by
Susan E. Thompson

Austrian Academy
of Sciences Press

Vienna 2008

Communications in Asteroseismology
Editor-in-Chief: **Michel Breger**, michel.breger@univie.ac.at
Editorial Assistant: **Daniela Klotz**, klotz@astro.univie.ac.at
Layout & Production Manager: **Paul Beck**, paul.beck@univie.ac.at

Institut für Astronomie der Universität Wien
Türkenschanzstraße 17, A - 1180 Wien, Austria
http://www.univie.ac.at/tops/CoAst/
Comm.Astro@univie.ac.at

Cover Illustration

The location of the workshop, Mount Cuba Observatory, Greenville, DE, and the Fourier window for the WET run XCOV25 on the DBV GD 358.

British Library Cataloguing in Publication data.
A Catalogue record for this book is available from the British Library.

ISBN 978-3-7001-6118-9
ISSN 1021-2043

Austrian Academy of Sciences Press
A-1011 Wien, Postfach 471, Postgasse 7/4
Tel. +43-1-515 81/DW 3402-3406, +43-1-512 9050
Fax +43-1-515 81/DW 3400
http://verlag.oeaw.ac.at, e-mail: verlag@oeaw.ac.at

Contents

The entire conference enjoying the sunshine. Back row: Harry Shipman, Steve Kawaler, Staszek Zola, Michael Montgomery, Fergal Mullally, Reed Riddle, Xiaojun Jiang, Scot Kleinman. Front Row: James Dalessio, Margit Paparo, S. O. Kepler, Hari Om Vats, Agnes Kim, Susan Thompson, Gerard Vauclair, Atsuko Nitta. Photographer: Judi Provencal

Comm. in Asteroseismology
Vol. 154, 2008

The Whole Earth Telescope Network's August 2008 Workshop

H. L. Shipman

Delaware Asteroseismic Research Center, Mt. Cuba Observatory and University of Delaware, Newark, DE 19716 USA

Introduction

The Whole Earth Telescope project hosted a meeting in August 2008 in Newark, Delaware. The purpose of this meeting was to review recent progress in the field, welcome new participants to the network, introduce new and old participants to the way that the new network functions, discuss how we operate, and make plans for the future. This paper summarizes the workshop as a whole. The following papers in this issue describe particular areas of progress.

A fair number of groups have put together networks of people and telescopes to concentrate on particular stars or particular areas of astronomy. Most of these networks have lasted for a few years and then fallen into disuse when the original project driving them was completed. In the field of variable stars, two networks have lasted for much longer: the delta Scuti network established by Michel Breger in the early 1980s and the Whole Earth Telescope, which had its beginnings in the late 1980s under the leadership of Ed Nather and Don Winget. Both of these networks still exist and are doing good science. See DARC (2007) for a more extended history of DARC and DSN (2005) for information on the Delta Scuti Network.

Starting in the early 1990s, the history of WET is described in a series of conference proceedings. The first WET workshop was held in Austin, Texas, in November 1991, and as far as I know there are no formal conference proceedings. There are proceedings for the second workshop (Meistas & Solheim 1993) held in Moletai, Lithuania; the third (Meistas & Solheim 1995), held in Ames, Iowa; the fourth (Meistas & Moskalik 1998), held in 1997 in Koninki, Poland; the fifth (Meistas & Vauclair 2000) held in 1999 in Bonas Castle, France, under NATO auspices.

The summer 2007 meeting was also the first meeting held since the relatively recent management change of the Whole Earth Telescope (WET) project.

In 2004, a variety of circumstances including funding difficulties led the WET leadership to encourage Judi Provencal and me to seek an alternative source of funding for WET. We obtained a major grant from the Crystal Trust. As a result, the Mt. Cuba Astronomical Observatory and the University of Delaware jointly established the Delaware Asteroseismic Research Center (DARC) which oversees a number of variable star projects. WET is by far the largest project running under the auspices of DARC.

This paper and the papers which follow in this special section of Communications on Asteroseismology is intended to give the flavor of the summer 2007 WET workshop. We decided not to make the effort to gather a "complete set of papers or to somehow include every comment or part of the discussion. Rather, we invited every workshop participant to send us a contribution if they wished. These papers can serve to give present and future readers of this journal a flavor for where the WET project is in 2008.

GD 358 and Testing Stellar Convection

An important paper published in the fall of 2004 stimulated an important and new line of thinking, one which has guided much of WET planning for the next few years. Mike Montgomery (2004) suggested that the nonlinearities in stellar light curves could be used as a way to test convection. Convection has been on various lists of unsolved problems in astrophysics (and physics) for several decades. We all use variants of the classical mixing-length theory because, really, there's nothing else sensible to do. Until Montgomery published these ideas in 2004, the only tests of the mixing-length theory were rather indirect.

The Montgomery scheme is conceptually quite simple and elegant. Pick a star whose pulsation spectrum is sufficiently complex that the pulsations are nonlinear, but sufficiently simple that a reasonable set of observations can permit its pulsation modes to be identified securely. The amplitude pattern of the pulsations will be established by the way that the star responds to these pulsations, the way that its temperature stratification changes in response to periodic changes in the total energy flow. The star's light curve will depend on how convection responds to the driving provided by the pulsations. Compare the actual light curve with models and you can test convection.

This challenge is ideally suited to the combination of telescopes which are currently available for WET. The WET network contains a large number of small telescopes where time is readily available. If a star is bright enough to be accessible to these telescopes, they can identify the pulsation modes. At the same time, a larger telescope (such as a 2-m or 4-m, or possibly even larger) can provide a light curve with high signal/noise, typically S/N>100, to test various theories of convection.

With these considerations in mind, we began to plan a WET run which eventually took place in May 2006. XCOV25 became the WET run which included one of the largest number of telescopes in the history of WET.[1] While our efforts to obtain data from really large telescopes such as the Hawaii 10-m and SOAR were not successful, 12 of the 20 telescopes involved in WET were 2-m or larger and thus were able to obtain the kind of detailed light curves that were needed to implement the Montgomery test.

The results from this WET run were presented a year after the run at two conferences (the white dwarf conference in Leicester and the variable star conference in Vienna), and in some papers in this volume. Nevertheless, the papers presented in the summer of 2006, a few months after the observations were obtained, presented much more than a simple look at the power spectrum. Those papers (see Provencal et al. 2007a, 2007b) demonstrate that the Montgomery test works, that the mixing length theory with a reasonable choice of parameters does indeed produce an adequate fit to the light curve. It has encouraged us to widen our search for more stars like GD 358 so that we can extend the Montgomery test. "Like GD 358" does not necessarily mean DBV pulsators; rather, it means stars of similar brightness, and similar types of pulsation spectra: complex enough to be nonlinear but not so complex as to make the analysis impossible.

The data set for GD 358 is extraordinarily rich, and as time goes on our interpretation becomes progressively more detailed. The papers presented in the summer of 2006 at the biennial white dwarf conference and the Vienna variable star conference (Provencal et al. 2007a, 2007b) contain much deeper interpretations of new kinds of phenomena than was the case before. There is yet more in the papers published in this journal. The final papers will be deeper still.

New Ways of Operating WET

With the re-establishment of the WET under Delaware auspices, we made some changes in the way it works. Some were dictated by the completion of the transition from a network of telescopes using the same Texas three-star photometers to a network of telescopes and CCD cameras. Some reflected the choices of the new leadership. Considerable amounts of time at the conference were spent discussing how well these choices had worked and how they could be improved.

The role of headquarters. In the past, a WET run always had a headquarters, because simply implementing the run was often more complex

[1] XCOV stands for extended coverage, and the number is the number of the WET run. A list of WET runs is given at the DARC website at http://www.physics.udel.edu /darc/wet/campaigns.htm.

than was the case for XCOV25. A small change in the technology which had surprisingly large implications was the availability of signals from atomic clocks over the internet. We no longer needed phone calls between sites in order to establish a uniform time base. As a result, emails sufficed in almost all cases to find out whether and when a particular site could observe, and when it could observe. Because there was only one target, no decisions had to be made.

Supporting Software. Headquarters became a data analysis center. Antonio Kanaan, who had developed a number of scripts which allowed us to use the power of IRAF to analyze large volumes of data quickly, led an effort to analyze the real data quickly. Judi Provencal, Andrezj Baran, and Mike Montgomery also participated in this effort. Even though there were about 100,000 images which comprised the final data set, the final light curve was basically in place at the end of the WET run. Data reduction has never been so speedy within WET. It has often been the choke point which has delayed the publication and analysis of data.

There was considerable discussion at the summer meeting regarding how we should handle the question of providing appropriate support to individual observers in order to make data acquisition and data reduction as easy as possible. Many telescopes and observers have their own approach to data acquisition; the only additional difficulty with WET is being sure to provide the appropriate information, especially timing, for those at headquarters. Then it is up to those at headquarters to use their favorite software for data analysis, which at the moment (early 2008) appears to be Antonio Kanaan's scripts. Delaware graduate student James Dalessio discussed some software he is developing which may make it easier to analyze data from stars with close companions and from sites where the image of the target star shifts position during the night. Reed Riddle talked about the capabilities of the XQED software package which had been used in XCOV23-24.

Transition to an all-CCD network. When WET was originally established, the technology of CCDs was relatively new and observing practices were changing rapidly. One of the founding principles of WET was that every telescope would have the same instrumentation. Now that CCDs are sufficiently well established, it has become neither practical nor sensible to insist that all sites use a standard instrument such as the Texas 3-star photometer. Prior to XCOV25, there was much discussion over the internet about what we should do. Many sites on XCOV25 used a specific instrument, namely a back-illuminated camera provided by Apogee. However, the use of Apogees was not universal. It has seemed that the different back-illuminated cameras are similar enough so that our data analysis for XCOV25 has worked.

Other issues. Scot Kleinman described his use of the PUBLISH software package which makes it possible to post light curves quickly to the internet site.

This will be tried in forthcoming WET runs and campaigns. The sense of the meeting was that failure to get light curves publicly available quickly was one important thing which should be improved after XCOV25.

After the meeting, it was clear that it was important to have periodic meetings such as this workshop. Issues large and small need to be discussed so that the Delaware group can take action to implement improvements. While email conversations and special get-togethers at meetings like the variable star meetings and the biennial white dwarf workshops are helpful, it seems that WET workshops are another important way that the WET team can move forward. The WET management council, known more informally as the Council of the Wise or COW, gave its approval to the way that the WET is being run, but made a strong suggestion that the bandwidth of the internet connection to the Mt. Cuba Observatory, the site of headquarters, be improved.

WET Science and Future WET Targets

The final selection of WET targets is done, as was done in the past, by the WET director (currently Judi Provencal), consulting with the COW and after considerable exchange of e-mails. However, one of the things that can be accomplished at a WET workshop can be a great deal of face-to-face discussion of the merits of various targets which were proposed for the next WET run, which is scheduled for April 2008. About half of the time at this workshop was devoted to such discussion.

It is probably not worth recording the details of this discussion in an overview of the workshop here. Target selection was made when most of this paper was written. Final editing of this paper took place after the April 2008 WET run took place, and this authors preliminary assessment is that the spring 2008 WET run, XCOV26, was a success. However, several points emerged from the discussion which are worth recording for wider circulation.

Kepler reviewed some results of his group and of the Texas group which have led to a considerable increase in the number of known variable white dwarf stars. He reported a number of 140, with 80 of them discovered as a result of the Sloan Digital Sky Survey. Unfortunately many of the newly discovered white dwarf variables are very faint. The dynamics of time availability on telescopes of various sizes is very much in flux right now, and one of the issues for the future of this field will be a changing definition of what is too faint for the community to work on. Kawaler pointed out that there is a senior review at the NSF of time-domain astrophysics going on as of the fall of 2007.

A novel concept which emerged from the discussion was the idea of designating one or more stars as targets for observation before such a star becomes a target for a full fledged WET run, where the 20-plus telescopes in the WET community are all brought to bear on one single target, where several observers are sent to remote places to operate telescopes, and where there is a full-fledged headquarters. Such precursor observations were actually done in calendar year 2007 for two stars: WD1524-0030 and G38-29. G38-29 is an example of a star where very little data existed beyond the discovery data from decades ago (1977).

After the workshop, WET director Provencal announced that there would be a mini-campaign on G38-29 in early November 2007. This almost became a WET run, though a tentative decision was made not to give it an XCOV designation. We did not have a "headquarters but Antonio Kanaan did come from Brazil to analyze the data. We did make the effort to send two observers to Hawaii. We obtained some 2-m observations of this star.

There was some discussion of future missions and large telescopes which might affect the WET community. Spacecraft like COROT and KEPLER will provide information on variable stars. These missions are often designed with brighter variables in mind, and time will tell how the white dwarf community can best take advantage of the capabilities given by these missions. The science working group of the ground-based LSST has contacted the white dwarf community for input, and again there will be some efforts to maintain contact.

We also heard some presentations from and about new partners. Xiaojun Jiang of the National Astronomical Observatory of China presented the capabilities of the observatories which currently exist in China, including both 1-m and 2-m telescopes. There are plans for additional telescopes in western China which may be extremely useful for WET science. He and his colleagues have been very active contributors to WET. Hari Om Vats presented a report on behalf of a new observational astronomy group at the Mt. Abu Observatory, near Gurushikhar, the highest peak in the whole of central India. They have a 1-m class telescope with a very good camera. This group participated in the November campaign on G 38-29. I presented a report on his negotiations with the Las Cumbres Observatory, which is planning to build a worldwide network of telescopes in the 1-2 meter range (see www.lcogt.net for details). While their primary purpose is observing planetary transits and gravitational lensing, they are also open to collaborations such as WET. Las Cumbres participated in the November campaign and the April WET run.

The community of astronomers who studies white dwarf stars has a tradition at its meetings, which is enthusiastically followed both at the meetings of the whole white dwarf community (which occur in even-numbered years) and in WET workshops. We have tours and we have very nice conference dinners. Science does get discussed at these tours, of course. WET projects coordinator Teresa Holton arranged an excellent tour of the nearby Winterthur house and gardens, which present the collections of P.S. DuPont, a man who assembled rooms of furniture to represent particular periods in American history. She also arranged an excellent dinner at Harry's Savoy Grill, a nearby restaurant which did an excellent job in presenting seafood characteristic of the Chesapeake Bay and parts of the coast which lie more to the northeast. We were pleased to host some of our donors at this dinner. Without the support of the Crystal Trust for the DARC and WET projects, none of this would have happened. The author of this paper also acknowledges support from the National Science Foundations Distinguished Teaching Scholars program.

References

DARC 2007. http://www.physics.udel.edu/darc/darc_history.html (Accessed January 14, 2008)

DSN 2005. http://www.univie.ac.at/tops/dsn/intro.html (Accessed April 24, 2008)

Meistas, E. G., & Solheim, J. E. 1993, Proceedings of the Second WET Workshop, BaltA, 2, 360

Meistas, E. G., & Solheim, J. E. 1995, Proceedings of the Third WET Workshop, BaltA, 4, 105

Meistas, E. G., & Moskalik, P. 1998, Proceedings of the Fourth WET Workshop, BaltA, 7, 1

Meistas, E. G., & Vauclair, G. 2000, Proceedings of the Fifth WET Workshop, BaltA, 9, 1

Montgomery. M. H. 2005, AJ, 633, 1142

Provencal, J., Shipman, H. L., & the WET Team 2007a, CoAst, 150, 293

Provencal, J. L., Shipman, H. L., Montgomery, M. H., et al. 2007b, 5th European Workshop on White Dwarfs ASP Conference Series, ASPC, 372, 623

Delicious seafood dinner at Harry's Savoy Grill.

A. Nitta at Winterthur Gardens.

Comm. in Asteroseismology
Vol. 154, 2008

Do You Need the WET: A Modest Proposal

S. J. Kleinman

Gemini Observatory, 670 N. A'ohoku Pl., Hilo HI USA 96720

The Whole Earth Telescope (WET) is an extremely powerful, but also extremely resource-demanding, tool. To help ensure the WET's enormous resources are called into play only when they are truly needed, I urge potential users to seriously consider other options to getting the necessary data before turning to the WET and I urge the WET to consider ways to establish a mechanism for obtaining large sets of single-site data for potential WET targets.

I want to raise one question which should be forefront in everyone's mind before an object is proposed or accepted as a WET target. *Does your object really need extended coverage?* The WET has become so well-established, so easy to use, that we often forget the immense amount of resources in time, money, and telescope access that a WET run costs. It is relatively easy to propose and observe a target with the WET these days. We've learned a lot since the early days and the WET runs very efficiently thanks to the dedicated hard work of the people at headquarters and elsewhere. It can seem quite simple, appropriate, and even logical to use the WET simply as a mechanism for obtaining lots of data on a target, and not necessarily as one only to obtain lots of *extended coverage* data on an object. Compounded by a growing list of new variables and new WET collaborators and science ideas, this already-existing temptation becomes even more pronounced.

What makes the WET unique are the quality and dedication of its people and its concentration on extended, 'round-the-globe coverage of its target stars. The utilization of these great strengths requires an immense amount of resource expenditure and we do a dis-service to the WET and its members when we call the WET into being for an observation that does not absolutely require its unique strengths and resources.

The WET enjoyed a large amount of very early success. The main reason being that the WET was born out of necessity. There was already a pre-existing backlog of extensive, yet unexplained, single-site data that we knew had to be supplemented by extended coverage data if we were to ever understand them. The existing single-site data nearly guaranteed that breaking the 1-day alias in our data would produce scientific discoveries. We knew both that the WET observations were going to give us new information that we had no other way of obtaining and that this information would lead us directly to new results.

Thanks to the large increase in the number of known white dwarf and other pulsators from the Sloan Digital Sky Survey (eg.Mukadam et al. 2004 and Nitta et al. 2005) and other efforts, we have a lot of new potentially-interesting targets to observe. However, being new discoveries, many of these objects lack significant single-, let alone multi-site data that justify the necessity of the WET. In addition, many of the new variables are dim ($g \approx 19, 20$ mag) and require 2–4m class telescopes to observe. This means there is more competition for telescope time than with 1–2m class objects and even more requires us to have solid background data to justify the use of a very large percentage of the world's 2–4m telescopes for a continuous 2-week period.

We therefore need to concentrate on using the WET *only* when its strengths are fully justified. To this end, we need to start gathering background single/multi-site data on these new variables to determine their suitability for the WET. With the WET taking up a couple weeks of telescope time a year on many of the suitable telescopes/instruments for our work, there aren't many options left to getting additional time for this background work.

One solution would be to obtain access to a dedicated, remote-operated, suitable telescope, perhaps something like the Las Cumbres network[1]. The WET provides ample opportunity for shared learning, education, and public relations, so the match up should be fairly easy to sell. Alternatively, or maybe in addition, instead of a single WET run concentrating on one or two objects, we could do a non-WET run using similar resources, but gathering data on many different objects during the run. We could even consider adjusting which telescope observes which objects in an effort to optimally affect alias patterns.

These are just two possible ideas to obtain necessary pre-WET data. There are certainly others, but what is clear to me, is that in the end, we must be able to prove before the WET is ever turned to an object that unique results are likely to follow. The WET is too costly a tool to employ when it is not absolutely necessary. We must always ask and answer the question: *Does this object really need the WET?*

References

Mukadam, A., Mullally, F., Nather, R. E., et al. 2004, ApJ, 607, 982

Nitta A., Kleinman, S. J., Krzesinski, J., et al. 2005, ASPC, 334, 585

---

[1] http://lcogt.net/

S. Kleinman enjoying the afternoon talks.

Comm. in Asteroseismology
Vol. 154, 2008

Preliminary Asteroseismology of EC20058-5234 and Limits on Plasmon Neutrinos

A. Bischoff-Kim

The University of Texas at Austin, Astronomy Department, 1 University Station, C1400, Austin, TX 78712, USA

Abstract

We present results from a preliminary asteroseismological analysis of the hot DBV EC20058-5234 based on an extensive grid of models. EC20058 is the only known stable DBV and also the hottest and provides an ideal laboratory to constrain the emission rate of plasmon neutrinos. In a Whole Earth Telescope (WET) campaign in 1997, Sullivan et al. (2008) have found 11 independent modes. We use the results of their work to perform a 6 parameter fit of the observed period spectrum. We find a best fit model consistent with the spectroscopically determined effective temperature and surface gravity for EC20058. We find that the periods can be fit successfully without invoking an uneven split of the 281s mode due to a magnetic field. Based on our best fit model, we compute rates of change for the two stable modes observed in the star, which in turn can be used to place tight constraints on plasmon neutrino emission.

Astrophysical Context

Above a temperature of about 26000 K, more than half of the luminosity of an average white dwarf comes from neutrino emission (Figure 1). In that temperature range, the neutrinos are mainly produced by the decay of photons coupled to a plasma (plasmons). If we measure a neutrino luminosity in hot white dwarfs, we are measuring plasmon neutrino rates. One can use the change in the pulsation periods over time to measure the neutrino luminosity. As a white dwarf cools, the period of a given mode increases because the interior is becoming less and less compressible. The faster the cooling, the faster the period increases. Mestel theory (Mestel 1952) predicts $\dot{P}$ if the white dwarf is leaking energy exclusively through photons. A higher $\dot{P}$ than expected means that the star is cooling faster than expected, and indicates an extra source of energy loss. $\dot{P}$ provides therefore a measure of the neutrino luminosity.

Following the discovery of pulsating DBs as hot as 28000 K, Winget et al. (2004) demonstrated that one could use hot DBVs to measure neutrino rates. EC20058-05234 is the only stable DBV we know and for now we must rely on measuring $\dot{P}$ for that star. EC20058 is also the hottest known DBV and as such, is a very good candidate for studying plasmon neutrino emission. Sullivan (2004) has collected over 10 years of data on EC20058 that can be used to determine the cooling rate of the star, though he has not published a $\dot{P}$ yet.

We present preliminary results from an asteroseismological study of EC20058 and make a prediction of how tightly we can constrain the plasmon neutrino emission rates given a measured $\dot{P}$ for EC20058. We begin with a brief introduction to plasmon neutrino physics before introducing EC20058's observed spectrum and what it tells us about the star's properties and the contribution of plasmon neutrino emission to its cooling.

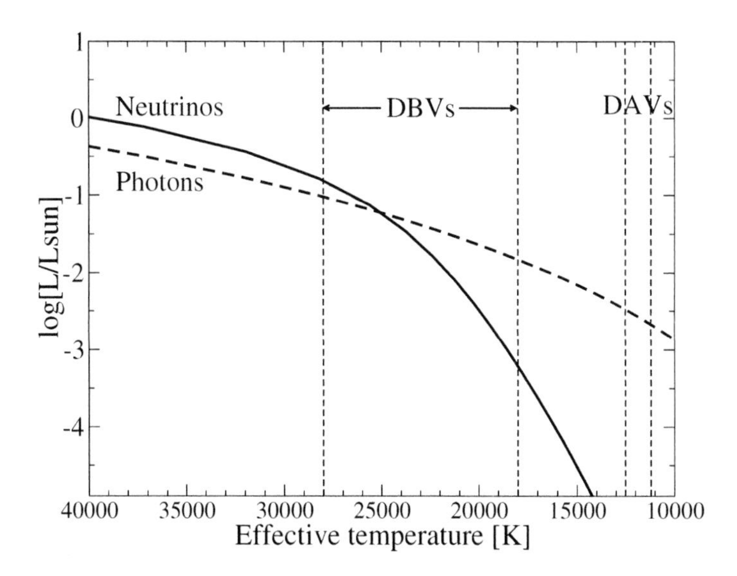

Figure 1: Time evolution of different sources of energy loss in a typical white dwarf model. The vertical dashed lines mark the location of the DBV and DAV instability strips. The neutrino luminosity remains significant near the hot (blue) edge of the DBV instability strip.

Plasmon neutrinos

Plasmon neutrinos result from the decay of a plasmon into a neutrino-antineutrino pair. Classically, one thinks of a plasmon as an electromagnetic wave propagating through a dielectric medium. It is made up of an oscillating electromagnetic field coupled with electrons oscillating with the same frequency. The frequency, ω, of such a wave obeys the dispersion relation

$$\omega^2 = \omega_0^2 + k^2c^2, \tag{1}$$

where ω_0 is the plasma frequency. Quantizing the field, the equation above yields

$$E^2 = \hbar^2\omega_0^2 + p^2c^2. \tag{2}$$

In effect, a plasmon is a particle similar to a photon, except it has a non-zero rest mass.

In free space, photons cannot decay into a neutrino-antineutrino pair without violating conservation of four-momentum. In a plasma, the electrons coupled to the photon allow conservation of energy and momentum and so plasmons can decay into a neutrino-antineutrino pair.

One can readily see from equation 1 that plasmons can only exist if $\omega > \omega_0$. This means that more plasmon neutrinos will be produced if the plasma has a high temperature relative to its frequency, i.e. $\mathrm{kT} > \hbar\omega_0$. However, ω should not be too large either because otherwise, the coupling with the electrons does not occur and plasmons do not exist. These conditions are satisfied in the interior of hot, neutrino emitting white dwarfs and is the dominant neutrino production process inside those stars.

Observed properties of EC20058-5234

EC20058 is the shortest period known DBV and the only one with stable pulsation periods. Spectroscopically, EC20058 appears to be hot and relatively low mass, with an effective temperature between 25600K and 29900K and a $\log(\mathrm{g})$ between 7.70 and 7.96 (Beauchamp et al. 1999). It is observable from the southern hemisphere and was first discovered to pulsate by Koen et al. (1995) from the South African Astronomical Observatory. In a 1997 Whole Earth Telescope (WET) run on this star, Sullivan et al. (2008) found 11 fundamental modes, listed in Table 1. The two highest amplitude modes f6 (281.0s) and f8 (256.9s) are remarkably stable and may be used to measure a cooling rate for EC20058 and constrain plasmon neutrino emission rates.

Sullivan et al. (2008) found two modes each split by 70 μHz. If we assume these modes are ℓ=1 modes (a reasonable assumption from asymptotic period

Table 1: Observed periods in EC20058

Mode name from Sullivan et al. (2008)	Period [s]	Notes
f1	539.8	
f2	525.4	
f3	350.6	
f4	333.5	
f5	(286.6)	m=-1 rotational split of f6
f6	281.0	High amplitude, stable mode
f7	274.7	possible m=+1 split of f6 (see text)
f8	256.9	High amplitude, stable mode
f16	(207.6)	m=-1 rotational split of f9
f9	204.6	
f11	195.0	

spacing arguments and the relatively low mass found from spectroscopy), then the 70 μHz splitting is consistent with a rotation period of the star of 2 hours. Sullivan et al. (2008) also suggest that perhaps f7 is the third member of the (f5,f6,f7) rotationally split ℓ=1 triplet, invoking the existence of a 3 kG magnetic field to account for the uneven frequency splitting. f7 could also be a mode of its own that happens to lie close to where the m=+1 member of the (f5,f6) multiplet would be if it were present. In the preliminary 6 parameter asteroseismological fits presented below, we consider both possibilities.

Asteroseismological Fits

In essence, the problem we have to solve in white dwarf asteroseismology is the simple minimization of a function (the average difference between the calculated periods and the observed periods) with n variables, including the stellar mass (M_*), the effective temperature (T_{eff}), and chosen structure parameters. Expressed mathematically:

$$\Phi(\mathrm{T_{eff}}, \mathrm{M}_*, ...) = \sigma_{\mathrm{rms}} = \frac{1}{\mathrm{N}} \Sigma_{\mathrm{i=1}}^{\mathrm{N}} \sqrt{\left(\mathrm{P_i^{calc}}\right)^2 - \left(\mathrm{P_i^{obs}}\right)^2}, \quad (3)$$

where N is the number of observed periods.

The simplest way to minimize Φ is to compute it for all conceivable values of the n variables and pick the smallest value we find. But the number of times we have to evaluate Φ can quickly become astronomical. To make matters worse, each evaluation of the function requires the full computation of a white dwarf model. Fortunately, the White Dwarf Evolution Code (WDEC,Lamb & Van Horn 1975, Wood 1990, see also Bischoff-Kim et al. 2008a for latest updates) allows us to compute a large number of models in a small amount of time on a standard desktop machine today and is perfect for the task.

To fit EC20058, we computed a grid of 286,416 models with simple core composition profiles and double-layered helium envelopes predicted by the time dependent diffusion of elements (Dehner & Kawaler 1995, Althaus et al. 2005). We show the core composition profiles of a representative model in Figure 2. The associated structure parameters are the central oxygen abundance (X_O), the location of the edge of the homegeneous carbon and oxygen core (q_{fm}), the location of the base of the helium rich layer (M_{env}), and the location of the base of the pure helium layer (M_{He}). Guided by the spectroscopy, we varied 6 parameters in the following ranges:

$$\begin{array}{ccccc} 25{,}600\ \mathrm{K} & \leq & T_{\mathrm{eff}} & \leq & 30{,}000\ \mathrm{K} \\ 0.45\ M_\odot & \leq & M_* & \leq & 0.57\ M_\odot \\ -2 & \leq & \log(M_{env}) & \leq & -4 \\ \log(M_{env})-2 & \leq & \log(M_{He}) & \leq & -8 \\ 0.5 & \leq & X_O & \leq & 1.0 \\ 0.35\ M_* & \leq & q_{fm} & \leq & 0.85\ M_* \end{array}$$

We tried fits including 8 periods, making the assumption that f7 in Table 1 is a member of the uneven triplet f5-f6-f7 due to a 2 hour rotation period of the star and a 3 kG magnetic field; and a 9 periods fit, not invoking a magnetic field and including f7 as a separate m=0 mode. We list the best fit found in the grid in both cases in Table 2.

Table 2: Best fit parameters for EC20058

Parameter	8 period fit	9 period fit
T_{eff}	28,000 K	28,400 K
M_*	0.54 $M_\odot$	0.54 $M_\odot$
$\log(M_{env})$	-3.6	-3.6
$\log(M_{He})$	-6.4	-6.4
X_O	1.00	0.80
q_{fm}	0.35 M_*	0.35 M_*
Φ	1.92s	2.20s

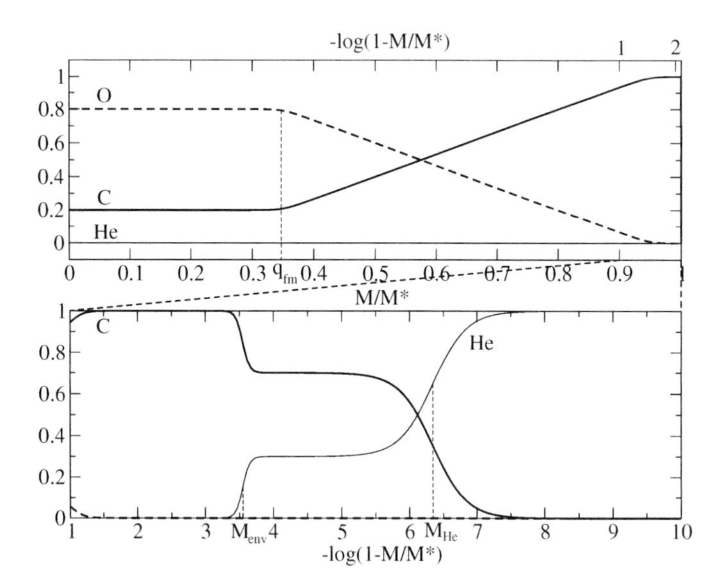

Figure 2: Core composition profiles for the grid models and internal structure parameters q_{fm}, M_{env}, M_{He}.

Both fits give similar answers and are equally good. While Φ for the 9 period fit is slightly higher, we have to compensate for the fact that we have added an extra period. Reduced χ^2's and the Bayes information criterion (BIC, Koen & Laney 2000) both indicate that the two fits are equivalent in quality. Occam's razor favors the 9 period fit (no need for a magnetic field in that case).

Potential Limits on Plasmon Neutrino Rates

We can now use the best fit model we found (for the 9 period fit) to calculate the rate at which the two stable modes (256.9 and 281.0s) change with time. We find $\dot{P}_{256.9} = 0.96 \times 10^{-13}$ s/s and $\dot{P}_{281.0} = 1.4 \times 10^{-13}$ s/s. In order to study the effect of varying the plasmon neutrino rates, we define the parameter λ as:

$$\epsilon'_\nu = \lambda \epsilon_\nu, \tag{4}$$

where ϵ_ν is the plasmon neutrino emission rates given by Itoh et al. (1996) and ϵ'_ν is the changed rate used in the models.

We show $\dot{P}$'s as a function of varying plasmon neutrino rates for both stable modes in Figure 3. In the upper right corner of the left panel, we show the error bars we expect on the measuremed $\dot{\mathrm{P}}$'s for these two modes. We estimate the error bars based on the ones we had in 1991 for G117-B15A (0.35×10^{-14} s/s, Kepler et al. 1991), after 15 years of data had been collected on that star.

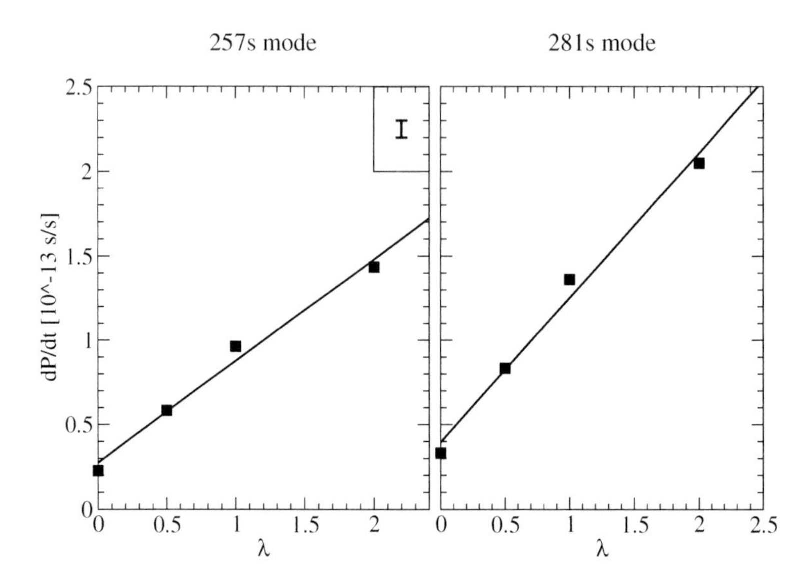

Figure 3: $\dot{\mathrm{P}}$'s as a function of λ (see equation 6 and text) for the two stable modes observed in EC20058; computed using the 9 period best fit model found in the previous section. In the box in the upper right corner of the left panel, we show the expected error bars on the measured $\dot{\mathrm{P}}$'s.

Conclusions

Preliminary asteroseismological fits based on an extensive grid of models give best fit parameters consistent with the spectroscopic temperature and gravity determination for EC20058 and indicate that the star is likely oxygen rich. We obtain good fits without invoking a magnetic field to explain the 274.7s mode observed in EC20058. From our best fit models we computed $\dot{\mathrm{P}}$'s for the plasmon neutrino rates predicted by Itoh et al. (1996), as well as for modified plasmon neutrino rates. We have set the stage to constrain plasmon neutrino rates from EC20058's $\dot{\mathrm{P}}$. Once we have a measured $\dot{\mathrm{P}}$ for one or both of EC20058's stable modes, we should be able to place tight constraints on plasmon neutrino emission rates. Most of the uncertainties lie in the models. More work is under way to improve the models and assess the associated uncertainties (Bischoff-Kim et al. 2008b).

Acknowledgments. This work was made faster and easier thanks to neatly packaged code provided by Dr. T. Metcalfe. This research was supported by NSF grant AST-0507639.

References

Althaus, L. G., Serenelli, A. M., Panei, J. A., et al. 2005, A&A, 435, 631

Beauchamp, A., Wesemael, F., Bergeron, P., et al. 1999, ApJ, 516, 887

Bischoff-Kim, A., Montgomery, M. H., & Winget, D. E. 2008a, ApJ, in press

Bischoff-Kim, A., Metcalfe, T. S., & Montgomery, M. H. 2008b, in preparation

Dehner, B. T., & Kawaler, S. D. 1995, ApJL, 445, L141

Itoh, N., Hayashi, H., Nishikawa, A., & Kohyama, Y. 1996, ApJS, 102, 411

Kepler, S. O., Winget, D. E., Nather, R. E., et al. 1991, ApJL, 378, L45

Koen, C., O'Donoghue, D., Stobie, R. S., et al. 1995, MNRAS, 277, 913

Koen, C., & Laney, D. 2000, MNRAS, 311, 636

Lamb, D. Q., & van Horn, H. M. 1975, ApJ, 200, 306

Mestel, L. 1952, MNRAS, 112, 583

Sullivan, D. J. 2004, ASP Conf. Ser. 310: IAU Colloq. 193: Variable Stars in the Local Group, 310, 212

Sullivan, D. J., Metcalfe, T. S., O'Donoghue, D., et al. 2008, ApJ, submitted

Winget, D. E., Sullivan, D. J., Metcalfe, T. S., et al. 2004, ApJL, 602, L109

Wood, M. A. 1990, Ph.D. Thesis

S. Kleinman, A. Nitta, R. Riddle and A. Kim-Bischoff gather around S. Kawaler's computer.

Comm. in Asteroseismology
Vol. 154, 2008

An Update on XCOV25: GD358

J. L. Provencal[1], H. L. Shipman[1], and The WET TEAM[2]

[1] Mt. Cuba Observatory and the University of Delaware
Dept. of Physics and Astronomy, Newark, DE 19716

[2] www.physics.udel.edu/darc/wet/

Abstract

We present a preliminary report on 436.1 hrs of nearly continuous high-speed photometry on the pulsating DB white dwarf GD358 acquired with the Whole Earth Telescope, in concert with the Delaware Asteroseismic Research Center (DARC) during May 12 to June 16, 2006.

Introduction

Asteroseismology of stellar remnants is traditionally thought of as the study of interior structure of pulsating white dwarfs and subdwarfs as revealed by global stellar oscillations. The light from stellar sources, be it detected using a 0.6m or a 10m telescope, originates from their surfaces. Stellar oscillations contain information about the interior of the star through which they have traveled, allowing a view beneath the photosphere. In the case of white dwarfs, the individual pulsations are temperature variations arising from nonradial g-modes. Asteroseismology allows us to retrieve information about basic physical parameters, including mass, rotation rate, and internal composition. This information (see for example: Winget et al. 1991, Winget et al. 1994, Kepler et al. 2003, Kanaan et al. 2005) provides important insights into a wide range of fields, from stellar formation and evolution, the chemical evolution in our galaxy, the age of the galactic disk, and the physics of Type Ia supernovae.

Asteroseismology is now expanding its focus to investigate problems that at first consideration may not be best suited for these techniques. Convection remains one of the largest sources of theoretical uncertainty in our understanding of stars. Our lack of understanding leads to considerable systematic theoretical uncertainties in such important quantities as the ages of massive stars (Di-Mauro et al. 2003) and the temperatures and cooling ages of white dwarfs (Wood 1992). Montgomery (2005) shows how precise observations of the light curves of variable stars can be used to characterize the convection zone in a particular star. Montgomery bases his approach on important analytical and numerical precursor calculations (Brickhill 1992, Goldreich & Wu 1999, Wu 2001). The method is based on three assumptions: 1) the flux perturbations are sinusoidal below the convection zone, 2) the pulsations can be treated to first order as if they were radial, and 3) the convective turnover time is short compared with the pulsations so the convection zone can be assumed to respond instantaneously. This approach can observationally determine the convective time scale τ_0, a temperature dependence parameter N, and, together with an independent T_{eff} determination, the classical convective efficiency parameter (the mixing length ratio) α.

In concert with the Delaware Asteroseismic Research Center (Provencal et al. 2005), we organized a WET run in May of 2006 with GD358 as the prime target (XCOV25). Our purpose was twofold: 1) obtain at least 5 hours of high signal to noise photometry from a large telescope and 2) accurately identify the frequencies, amplitudes and phases present in GD358's current pulsation spectrum. We fulfilled both of our goals. In the following, we will provide a preliminary overview of the data set and reduction procedures and present a preliminary list of identified modes, combination frequencies, and multiplet structure.

The Observations

XCOV 25 spans May 12 to June 14. Nineteen observatories participated in the run, contributing a total of 88 runs (a complete list of participants and observing runs can be found at www.physics.udel.edu/darc/wet/XCov25/xcov25.html). We obtained 436.1 hrs of observations, achieving 73% coverage during the main portion of the run.

Recent WET runs (examples include Kanaan et al. 2005) comprise a mixture of CCD and PMT observations, and XCOV25 is no exception. CCDs were employed at sixteen observatories, and 3-channel PMT photometers at the remaining sites. We attempted to minimize bandpass issues by using CCDs with similar chips and equipping each CCD with a BG40 or S8612 filter to normalize wavelength response and reduce extinction effects. The bi-alkali photomultiplier tubes are blue sensitive, so no filters were required. We also made every attempt to observe the same comparison star at each site.

Standard procedure for a WET run is for observers to transfer observations to headquarters for analysis at the end of each night. In the past, CCD observers completed initial reductions (bias, flat, and dark removal) at their individual sites, performed preliminary photometry, and transferred the result to WET headquarters. For XCOV25, the majority of participants using CCD photometers transferred their raw images, enabling headquarters to funnel data through a uniform reduction pipeline. The few sites unable to transfer images nightly performed preliminary reductions on site using the same procedures as those at headquarters, and sent their images at a later date.

The PMT data were reduced using the WET standard prescription developed by Nather et al. (1990). CCD data reduction followed the pipeline described by Kanaan et al. (2002). Figure 1 presents the lightcurve from the central 2 weeks of XCOV25 observations.

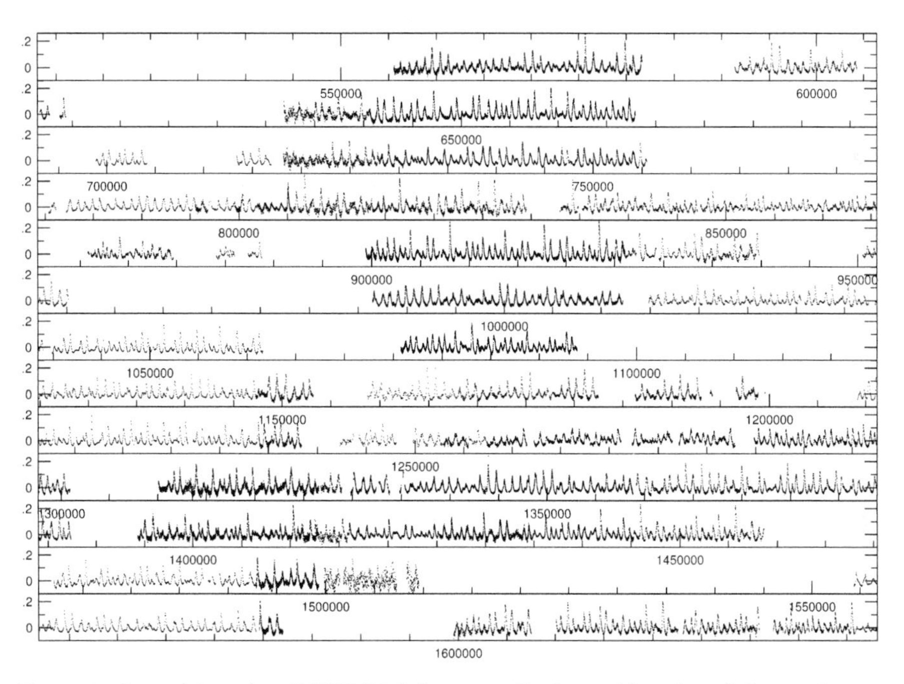

Figure 1: Central 2 weeks of XCOV25 light curve. Each panel is 1 day of observations.

Figure 2 presents the Fourier Transform (FT) of the entire data set. We carried out multi-frequency analysis using the Period04 software package (Lentz 2004). The basic method involves identifying the largest amplitude peak in the FT, subtracting that sinusoid from the original lightcurve, recomputing the FT, examining the residuals, and repeating the process. This technique is fraught with peril, as it is possible for overlapping spectral windows to conspire to produce alias amplitudes larger than any real signal. Our final identifications result from a simultaneous nonlinear least squares fit of 130 frequencies, amplitudes, and phases, some of which are labelled in Figure 2. We employed this procedure to identify 130 frequencies satisfying our criteria of amplitudes 4σ above the average noise.

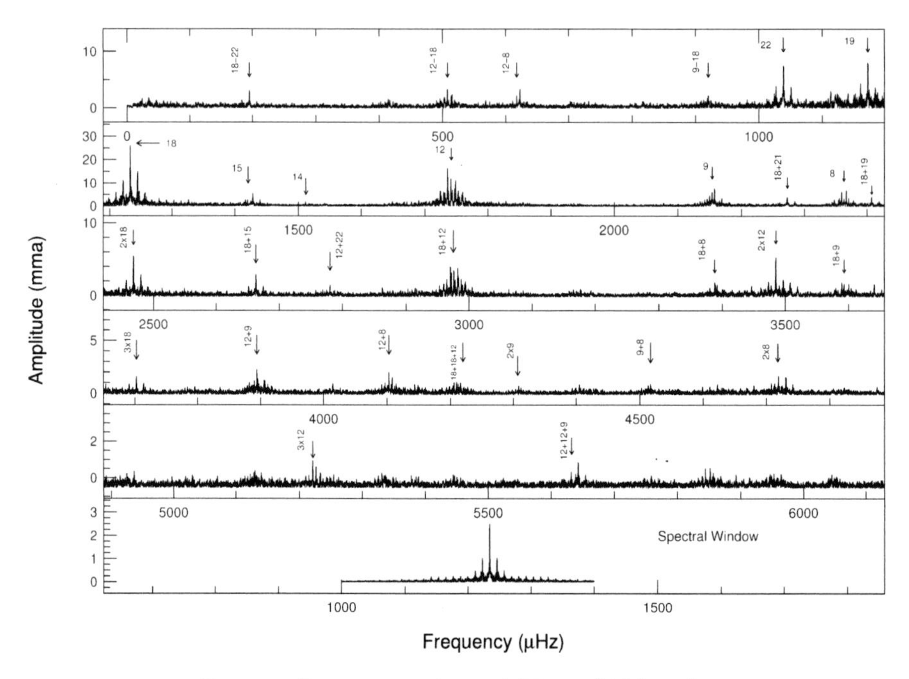

Figure 2: Fourier Transform of GD358 (XCOV25)

The data set contains a significant fraction of overlapping data. We experimented with the effects of overlapping data on the FT by computing FTs with 1) all data included, 2) no overlapping data, where we kept those data with higher signal to noise ratio, and 3) weighting the overlapping lightcurves by telescope aperture size. Figure 3 presents a comparison between the spectral window of the entire run and the central 14 days shown in Figure 1. We found no significant differences between the spectral window or FTs of overlapping versus nonoverlapping versus weighted data.

The Fourier Transform

Table 1 lists a sample of preliminary frequency identifications, due to space considerations. The complete list can be found in Provencal et al. (2008). We adopt the identifications of Winget et al. (1994), but must emphasize that

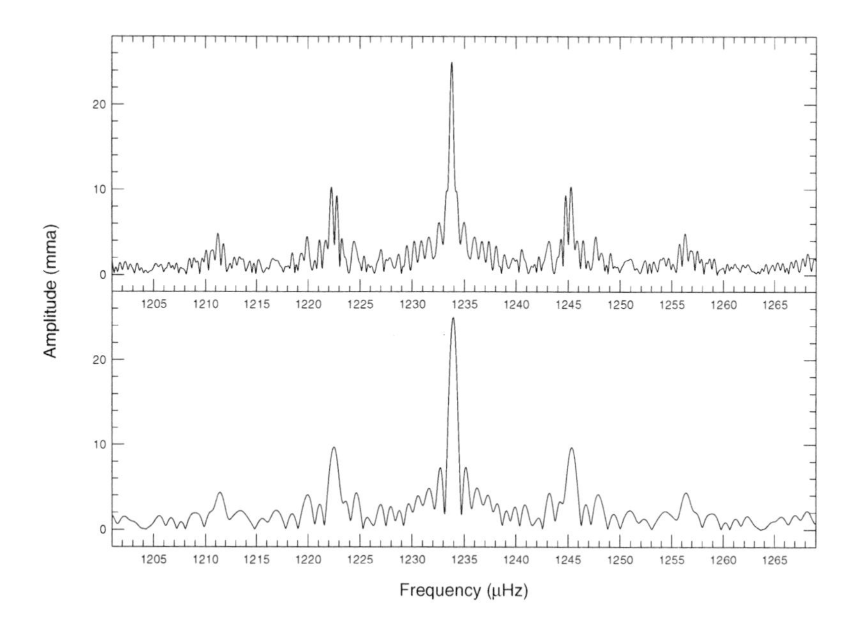

Figure 3: Comparison of the spectral window of the entire WET run with that from the main portion of the run given in Figure 1.

while we are confident from previous work that the dominant modes are l=1, the actual k identification cannot be observationally determined and may not correspond precisely to the values given here.

The dominant mode in 2006 is k=18 (1234.124 μHz, 810.291 s) with an average amplitude of 24.04 mma. Mode 18 is detected in previous observations (see Kepler et al. 2003) but not as the dominant frequency. The previously dominant k=15 and 17 modes have greatly diminished amplitudes, and we do not detect k=16 or 13. Perhaps the greatest surprise is the appearance of prominent power near the predicted value for k=12, a region of the FT previously devoid of significant peaks. This mode was detected in 1990, 1994, 1996, and 2000 but never at an amplitude about 1 mma. We do not detect the suspected l=2 mode at 1255.4 μHz noted in Kepler et al. (2003). We also do not find k=7 at 2675.5 μHz. (Kepler et al. 2003) suggest that this mode may have been excited to visibility via resonant coupling with k=17 and 16. Since

k=17 and 16 do not have significant amplitude in 2006, it follows that k=7 would not be detected.

GD358's FT contains a rich distribution of combination frequencies, from simple harmonics to fourth order combinations, some of which are given in Table 1. Most of these combinations are exact to within statistical uncertainties. Combination peaks, whose frequencies are linear combinations (both sums and differences) of 2 or more mode components, are typically observed in large amplitude pulsators (Dolez et al. 2006, Thompson et al. 2003). The general consensus on the origin of these peaks (Brassard et al. 1995, Yeates et al. 2005) argues that they are indicative of nonlinear distortions induced by the propagating medium, in this case the convection zone. The convection zone acts as a nonlinear filter, varying its depth in response to the pulsations, and distorting the original sinusoidal variations. We expect the amplitudes of the observed combinations, and the number associated with a given parent mode, to be a function of the flux intensity of the mode(s) involved.

Comparison with Other Observing Seasons

Figure 4 presents a sampling of GD358's FT for seasons spanning 1990 to 2006. We pause here to reflect on our previous mention of spectral windows and alias patterns. The 1990, 1991, 1994, 2000, and 2006 FTs are from WET runs, and the other seasons are single site, obtained from McDonald Observatory. Comparison of the FTs and corresponding spectral windows dramatically illustrates the power of WET to reduce alias artifacts.

A simple visual comparison of the FTs in Figure 4 illustrates the simultaneous simplicity and complexity of GD358. While large amplitude peaks are always confined between 1000-1800 μHz, and individual modes appear in the same general location over the years, the distribution and amplitude of excited modes varies. Figure 5 shows the frequency of the largest amplitude peak in each FT, from 1982 to 2007.

Table 1: A Sample of Identified Frequencies

FrequencyμHz $\pm 0.001\mu$Hz	Amplitude (mma) $\pm 0.07 mma$	Note
195.685	2.70	18-21
617.431	2.03	18/2
1039.076	7.94	k=21
1173.015	7.24	k=19
1222.946	4.30	k=18
1228.792	5.06	k=18
1234.124	24.03	k=18
1239.511	4.93	k=18
1245.220	4.90	k=18
1429.210	5.63	k=15
1512.141	1.80	k=14
1736.311	16.35	k=12
1737.962	5.60	k=12
1741.666	11.01	k=12
1743.738	5.60	k=12
1746.672	1.81	k=12
1749.083	10.92	k=12
1856.845	1.41	k=11
2150.393	4.10	k=9
2154.224	5.51	k=9
2158.074	7.18	k=9
2273.691	4.23	18+21
2359.053	5.95	k=8
2363.058	1.64	k=8
2366.524	6.60	k=8
2407.205	3.80	18+19
2468.282	5.19	2x18
2663.368	2.95	18+15
2909.416	1.00	18+12
2964.917	1.10	18+12
2970.400	3.01	18+12
2972.085	2.82	18+12
2975.814	3.47	18+12
2977.885	1.71	18+12
2981.032	1.31	18+12
2981.947	3.53	18+12
2983.266	3.12	18+12
2988.643	1.10	18+12

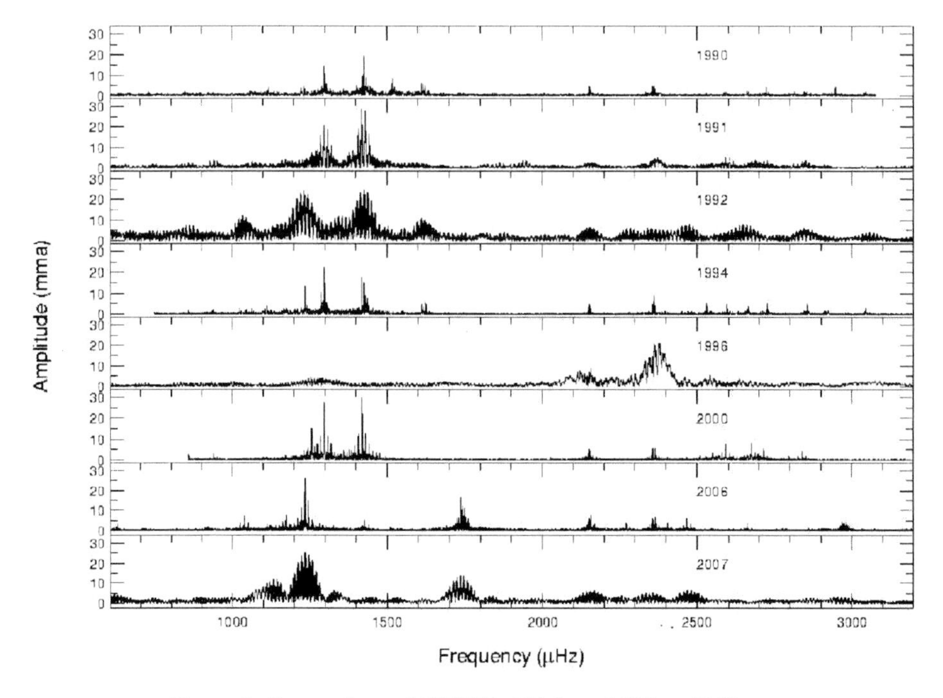

Figure 4: Comparison of GD358's FT from 1990 to 2006

Multiplet Structure

The analysis of fine structure multiplets in pulsating white dwarfs is based on the assumption that the multiplets are produced by lifting of the degeneracy of the azimuthal quantum number m by rotation and/or magnetic fields. In the limit of slow rotation, we expect the observed fine structure to reflect the star's rotation rate and the spherical harmonic degree l of the pulsation involved, with possible perturbations introduced by any surface magnetic field. We also expect the fine structure to remain stable over long time periods. A classic example is the prototype DO pulsator PG1159-035, which exhibits beautiful triplets, corresponding to l=1, and quintuplets, corresponding to l=2 (see Figures 5 and 6 in Winget et al. 1991). All of the multiplets of a given l have the same frequency splitting, with ratio of different l values very close to the expected theoretical prediction.

Figure 6 presents a "snapshot" of multiplet structure in the XCOV25 FT. Winget et al. (1994) (hereafter W94) identified triplet structure for most modes, with frequency splittings that varied with both k and m. In 2006, the only modes exhibiting clear triplet structure with splittings in agreement

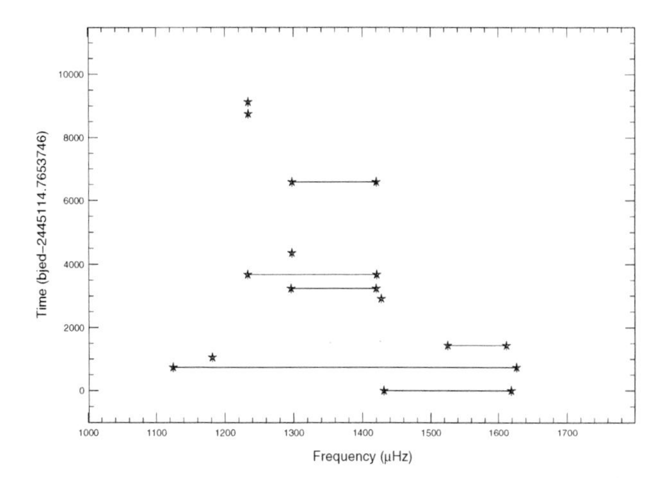

Figure 5: Frequency of the largest amplitude peak in GD358 from 1982-2007. The y axis shows time, and the x-axis displays frequency. The time is given in baryocentric julian ephemeris date. Tzero is 1982. In several seasons, there are two peaks that have amplitudes within 2 mma. We have included both frequencies and joined them with a line for clarity.

with previous observations are k=9 and 8. We find average multiplet splittings of 3.83 and 3.75 μHz, respectively. The only other mode we detect in common with both Winget et al. (1994) and Kepler et al. (2003) that has sufficient amplitude to investigate fine structure is k=15, but the multiplet structure is quite different from previous reports. In 1990 (Figure 7), multiplet 15 had a reported average multiplet splitting of 6.4 μHz. In 1994, the value was $\approx 6.7\mu$Hz, and in 2000, $\approx 6\mu$Hz. In 2006, k=15 contains multiple components with a dominant splitting of $\approx 5.4\mu$Hz. The 5.4 μHz splitting also appears in k=18 (a quintuplet in 2006) and k=*12*. We point out that the other high k modes (17, 16, 14, and 13) reported by W94 to have frequency splittings of $\approx 6\mu$Hz are not detected here.

Conclusions and Speculations

The analysis of this data set is still underway, but has already provided new insight into GD358. As is true with any new observations, the data set has

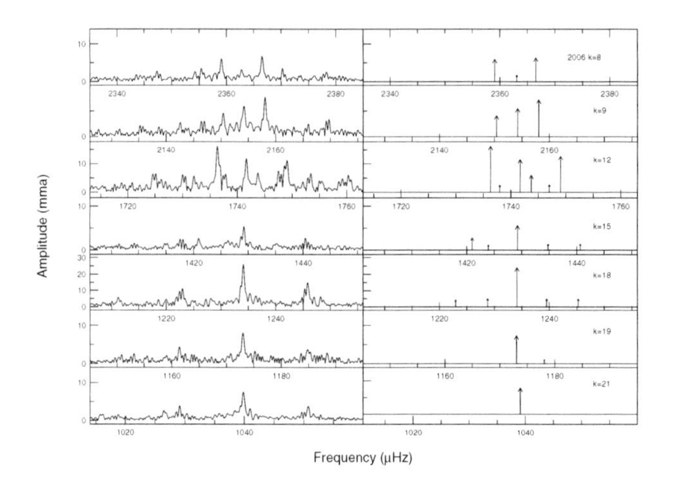

Figure 6: A "snapshot" of modes and their multiplet structure for 2006. The left panels give the actual FTs, and the right panels give the prewhitening results. Each panel spans 50 μHz. Note the change in y-scale.

also raised many new questions. The role of the convection zone in nonlinear pulsators is becoming clearer. For example, convection does not play a role in the DOV pulsators, and the prototype PG1159-035's FT displays expected triplets and quintuplets corresponding to l=1 and 2 pulsations, and does not contain combination frequencies. Convection does play a role in GD358, and we find no evidence for stable multiplet structure here, and we do find a myriad of combination frequencies. Reviving memories of basic physics demonstrations, water in a tank will reflect off the tank walls. In a star, the bottom of the convection zone plays the role of the wall. Yet, because the star is pulsating, the convection zone is constantly changing. For an m=0 mode, the poles appear to recede, but the equator does not. In other words, the convection zone does not always form a perfectly spherical reflective surface. Could this explain the difference in behavior of the various modes in GD358? Can this explain the apparent changes in mode amplitudes we observe?

We are also investigating the possible role of magnetic fields. As a simple model, one could imagine switching on a global magnetic field on a rotating, pulsating white dwarf. Our imaginary magnetic field is confined to the non-degenerate atmosphere of the white dwarf. Its influence would be strongest at

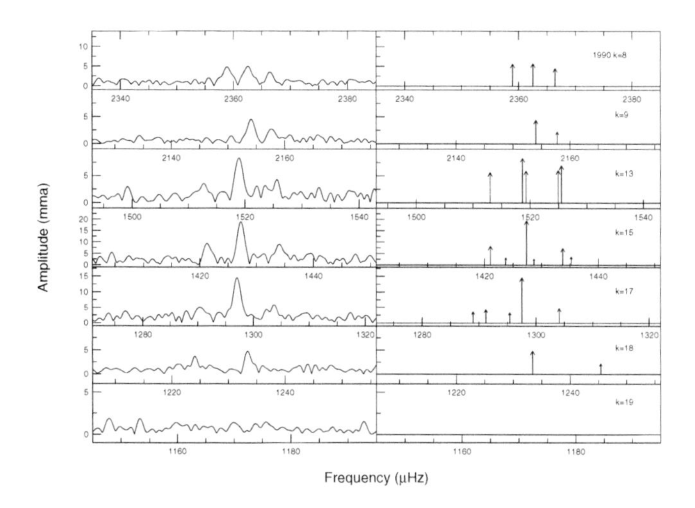

Figure 7: A "snapshot" of modes and their multiplet structure for 1990. The left panels give the actual FTs, and the right panels given the prewhitening results. Each panel spans 50 μHz. Note any changes in y-scale.

the surface where the gas pressure is decreasing, but not strong enough to affect rotation or convection at the surface. If the field is a dipole, each pulsation mode could be split into $(2l+1)^2$ components. For l=1, each mode could contain up to 9 components. Since the high k modes preferentially sample the surface, could the presence of a non-aligned, variable magnetic field explain the dramatic changes in multiplet structure we observe, while the low k modes, which are not influenced as strongly by the magnetic field and/or the convection zone, are left relatively unaffected?

Our continuing investigation into these questions includes a detailed analysis of multiplet structure over time, a closer look at the combination frequencies, an effort to detect a possible magnetic cycle period, and continued monitoring of GD358 and other high amplitude pulsators.

Acknowledgments. DARC acknowledges the support of the Crystal Trust Foundation and Mt. Cuba Observatory. We would also like to thank everyone involved in the network for their time and support in obtaining these observations.

References

Brassard, P., Fontaine, G., & Wesemael, F. 1995, ApJS, 96, 545

Brickhill, A. J. 1992, MNRAS, 259, 519

Goldreich, P., & Wu, Y. 1999, ApJ, 511, 904

Kanaan, A., Kepler, S. O., & Winget, D. E. 2002, A&A, 389, 896

Kanaan, A., Nitta, A., Winget, D. E., et al. 2005, A&A, 432, 219

Kepler, S. O., Nather, R. E., Winget, D. E., et al. 2003, A&A, 401, 639

Lenz, P., & Breger, M. 2004, IAU Symposium, 224, 786

Nather, R. E., Winget, D. E., Clemens, J. C., et al. 1990, ApJ, 361, 309

Montgomery, M. 2005, ApJ, 633, 1142

Provencal, J. L., Shipman, H. L., & the Wet Team 2005, BAAS, 37, 1157

Provencal, J. L., et al. 2008, in preparation

Thompson, S. E., Clemens, J. C., van Kerkwijk, M. H., et al. 2003, ApJ, 589, 921

Winget, D. E., Nather, R. E., Clemens, J. C., et al. 1991, ApJ, 378, 326

Winget, D. E., Nather, R. E., Clemens, J. C., et al. 1994, ApJ, 430, 839

Wu, Y. 2001, MNRAS, 323, 248

Yeates, C. M., Clemens, J. C., Thompson, S. E., & Mullally, F. 2005, ApJ, 635, 1239

J. Provencal, the WET's director, enjoying herself at Winterthur.

Comm. in Asteroseismology
Vol. 154, 2008

What We Can Learn from the Light Curves of GD 358 and PG 1351+489

M. H. Montgomery

The University of Texas at Austin, Department of Astronomy, 1 University Station, C1400, Austin, TX 78712, USA
Delaware Asteroseismic Research Center, Mt. Cuba Observatory, Greenville, DE 19807, USA

Abstract

We present the results of light curve fits to two DBV white dwarfs, PG 1351+489 and GD 358. For both fits, we include recent improvements in calculations relating the bolometric and observed flux variations. We provide a preliminary map of convection across the DB instability strip, and we show how this allows us to choose between two possible spectroscopic fits to PG 1351+489.

Astrophysical Context

White dwarf stars come in two basic varieties: those with nearly pure helium surface layers (DBs) and those with nearly pure hydrogen surface layers (DAs). Each of these types contains a subtype of pulsators, denoted by DBV and DAV, respectively. In fact, the DBV GD 358 is the original member of the class of DBVs: it was discovered and predicted to pulsate by Winget (Winget 1982; Winget et al. 1982) and, as such, it is probably the most extensively studied DBV.

Since $\sim 98\%$ of all stars become white dwarfs, white dwarfs hold the key to understanding late stages of stellar evolution. In particular, the pulsators, through their observed oscillation frequencies, provide us with detailed information about their internal structure. In this regard, GD 358 is again one of the most successful examples. Asteroseismological fits to its frequencies have yielded constraints on not just is mass, $T_{\rm eff}$, and rotation rate, but also its internal composition and chemical profiles (Metcalfe, Nather, & Winget 2000; Bradley & Winget 1994; Winget et al. 1994).

In addition to the main pulsation frequencies, these pulsators contain information which has not been traditionally exploited: the *non*-linearities in the light curves of medium and large amplitude pulsators. Some work has been done in this area. It was Brickhill (1992) who originally showed the convection zone should produce the strongest nonlinearities in the light curves, and this was revisited by Wu (2001), who gave an analytical basis to this approach. Further, Yeates et al. (2005) applied Wu's approach to the problem of mode identification in DAV stars. Finally, we have recently developed a technique for directly modelling these light curves which makes full use of the nonlinearities present in the pulsations (Montgomery 2005, 2007a,b). This approach removes some of the parameter degeneracies of the second-order analytical expansion of Wu, and has so far led to mode identifications in two stars: the DAV G 29-38 and the DBV PG 1351+489.

The Light Curve Model

We make the same set of physical assumptions as given in section 2.1 of Montgomery (2005), i.e.,

1. The flux perturbations beneath the convection zone are sinusoidal in time and have the angular dependence of a spherical harmonic.

2. The convection zone is so thin that we may locally ignore the angular variation of the nonradial pulsations, i.e., we treat the pulsations locally as if they were radial.

3. The convective turnover timescale is so short compared to the pulsation periods that the convection zone can be taken to respond "instantaneously".

4. Only flux and temperature variations are considered, i.e., the large-scale fluid motions associated with the pulsations are ignored.

Simple mixing length theories of convection predict that the thermal response timescale of the convection zone, τ_C, should be

$$\tau_C \approx \tau_0 \left(\frac{T_{\rm eff}}{T_0}\right)^{-N}, \tag{1}$$

where $T_{\rm eff}$ is the *instantaneous* effective temperature, T_0 is its time average, and τ_0 is the average value of τ_C. The exponent N has a value of ~ 90 for the DAVs and ~ 25 for the DBVs. This high power of $T_{\rm eff}$ means that other nonlinear processes may well be negligible in comparison. These assumptions lead to the following equation relating the fluxes:

$$F_{\rm ph} = F_{\rm base} + \tau_C \frac{dF_{\rm ph}}{dt}, \tag{2}$$

where $F_{\rm base}$ is the flux incident at the base of the convection zone, $F_{\rm ph}$ is the flux which emerges from the top of the convection zone in the photosphere, and τ_C is the *instantaneous* thermal response timescale, which is a function of $T_{\rm eff}$ and therefore $F_{\rm ph}$.

We have also made some important technical improvements to the light curve fitting code. First, we have extended it to the case where many modes having different ℓ and m values are simultaneously present, i.e., the flux at the base of the convection zone is now given by a sum over the number of modes:

$$\frac{\delta F_{\rm base}}{F_{\rm base}} = {\rm Re}\left\{\sum_{j=1}^{M} A_j e^{i\omega_j t+\phi_j} Y_{\ell_j m_j}(\theta,\phi)\right\}. \tag{3}$$

In this formula, A_j, ω_j, ϕ_j, ℓ_j, and m_j are the amplitude, angular frequency, phase, ℓ, and m values of the j-th mode, and the total number of modes is M.

Second, we have adapted the code to simultaneously fit an arbitrary number of observations ("runs"). This was a necessary step for applying it to the multiple runs obtained during an observing campaign. Since memory is allocated and deallocated as needed, the code typically uses only 8 MB of RAM, independent of the number of runs being fit.

Finally, we have improved the way in which we calculate the "flux correction" which is needed to convert bolometric flux variations into the variations obtained in the observed passband. Specifically, if we denote by F_X the flux in the passband X, we need an estimate of the quantity α_X defined by

$$\frac{\delta F_X}{F_X} = \alpha_X \frac{\delta F_{\rm bol}}{F_{\rm bol}}, \tag{4}$$

where $F_{\rm bol}$ is the bolometric flux, and δF is the variation in the flux due to the pulsations. Clearly, this factor depends on the wavelength coverage of the passband. Assuming an average wavelength response of ~ 5000 Å, in previous analyses I estimated that $\alpha \sim 0.42$ for DBVs and $\alpha \sim 0.66$ for DAVs. These values are not that different from what one obtains by assuming a blackbody law at temperatures appropriate for these objects. For the DBVs, however, these values *are* significantly different from what one expects from detailed model atmospheres.

In order to better calculate α_X, I employ grids of photometric indices tabulated by P. Bergeron[1] (for a detailed description see Bergeron, Wesemael, &

[1] http://www.astro.umontreal.ca/~bergeron/CoolingModels/

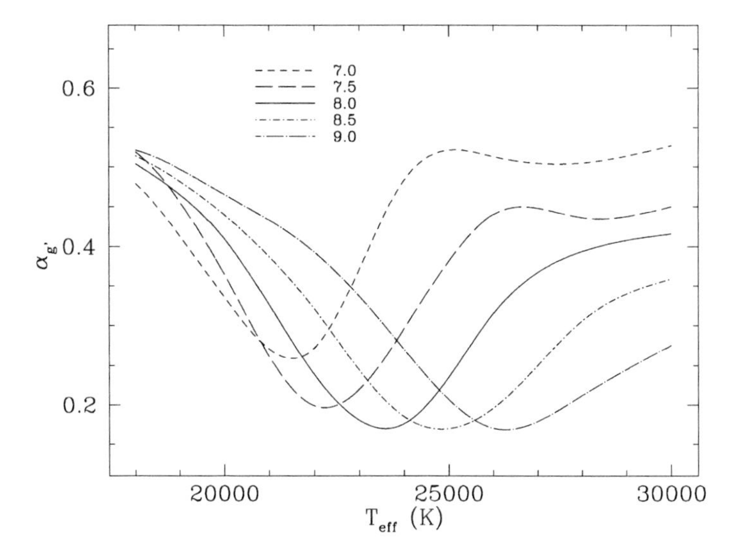

Figure 1: The flux factor $\alpha_{g'}$ as a function of $T_{\rm eff}$. The legend shows which values of $\log g$ correspond to the different curves.

Beauchamp 1995; Holberg & Bergeron 2006). These tables provide absolute fluxes in different passbands (e.g., Johnson U, B, and V) as a function of $T_{\rm eff}$ and $\log g$. Since flux changes are due almost entirely to temperature variations, we can rewrite equation 4 as

$$\begin{aligned} \alpha_X &= \frac{d\ln F_X}{d\ln F_{\rm bol}} = \frac{1}{4}\frac{d\ln F_X}{d\ln T_{\rm eff}} = -\frac{\ln 10}{10}\,\frac{dM_X}{d\ln T_{\rm eff}} \\ &\approx -0.230\,\frac{dM_X}{d\ln T_{\rm eff}}, \end{aligned} \tag{5}$$

where M_X is the absolute magnitude in the passband X. Our recent CCD observations have been made using a BG40 filter having an average wavelength response just below 5000 Å, which makes it a good match for the SDSS g′ filter. Thus, we take X to be g′ for the CCD data. Phototubes are more blue sensitive, with an average response of ~ 4200 Å, which is roughly equivalent to the Johnson B filter. Thus, for data taken with phototubes we use $X =$ B.

In Figure 1 we show $\alpha_{g'}$ as a function of $T_{\rm eff}$, for several different values of $\log g$. For $\log g = 8.0$ we clearly see that $\alpha_{g'}$ is typically less than the value of 0.42 assumed in our previous analyses, and can be up to twice as small depending on the temperature range. This figure also makes clear that taking $\alpha_{g'}$ to be a constant is not necessarily adequate for temperature excursions of several thousand degrees. Nevertheless, we defer an examination of this effect to a future publication and we simply assume that $\alpha_{g'}$ is constant during the pulsations.

The Fits

Previous to the 2006 WET run on GD 358, nonlinear light curve fits had only been made for mono-periodic pulsators (the DAV G 29-38 and the DBV PG 1351+489). This was because 1) the data can be folded at the period of the dominant mode, producing a high S/N "light curve", and 2) the number of possible mode identifications (ℓ and m values) for a single mode is small enough that all possibilities can be directly explored.

GD 358

The first *multi*-periodic pulsator to be explored, GD 358, violates both of the these conditions. First, due to its large number of large amplitude modes, the pulse shape obtained by folding its light curve at a mode period will not be the same as the pulse shape which would be obtained in the absence of other modes (Montgomery, 2007b). Second, due to the large number of excited modes it is not possible to search for all possible ℓ and m values for each mode. For instance, if we take it to have of order ~ 10 modes observed, all of which have $\ell = 1$, that yields a total number of cases of $(2\ell+1)^{10} \sim 60000$! Since each fit takes of order an hour, this is completely impractical using a standard desktop computing approach. However, T. Metcalfe (personal communication) points out that this problem could run in a few hours on a computer cluster having of order one thousand nodes.

Fortunately, GD 358 has been well studied, so we believe we have a good idea what the ℓ and m values for the main pulsation modes are (Metcalfe, Nather, & Winget 2000; Winget et al. 1994). Furthermore, from the 2006 WET run we obtained very good frequencies for these modes. Thus, since we can assume the frequencies and mode identifications to be known, we can make nonlinear light curve fits to individual observing runs within the WET campaign which have high S/N. High S/N data is necessary since we are mainly interested in the nonlinear part of the light curve, which itself is smaller than the linear part.

The highest S/N data taken during the campaign were obtained with the 2.7m Nordic Telescope (NOT) in La Palma. In Figure 2 we show the last night of this data together with the fit (this was a simultaneous fit of the last four night's of data taken by the NOT, of which we are showing only the last night). While not perfect, the fit clearly does a good job of reproducing most of the features in the light curve.

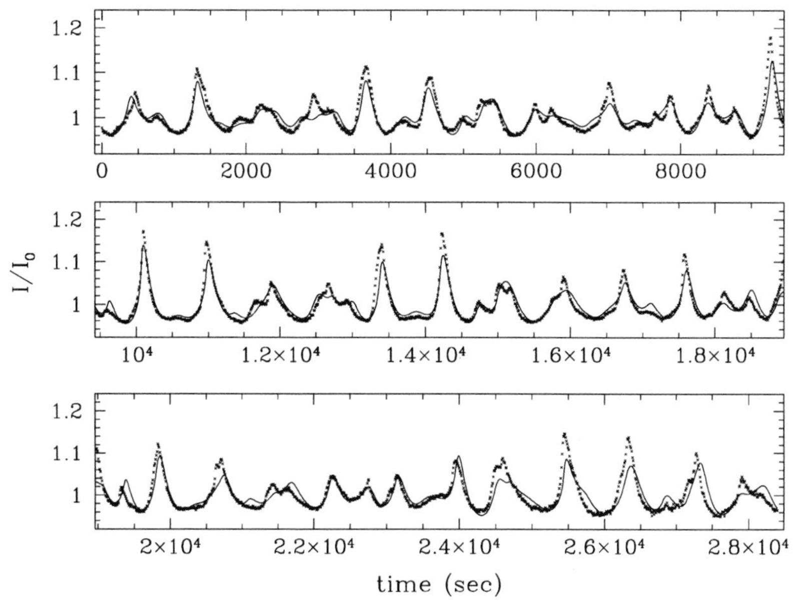

Figure 2: A light curve fit to data taken on GD 358 in the May/June 2006 WET campaign with the NOT 2.7m telescope. The data points are shown as crosses and the fit is the solid curve.

Table 1: Results of 12-mode Fit for GD 358.

Fit parameters:	$\tau_0 = 308$ s, $N = 22.2$, $\theta_{\rm i} = 46.6°$		
Period (s)	ℓ	m	Amplitude
422.561	1	1	0.17162
423.898	1	-1	0.13232
463.376	1	1	0.23350
464.209	1	0	0.10113
465.034	1	-1	0.10057
571.735	1	1	0.25407
574.162	1	0	0.17763
575.933	1	-1	0.32532
699.681	1	0	0.07525
810.291	1	0	0.35596
852.502	1	0	0.13170
962.385	1	0	0.12794

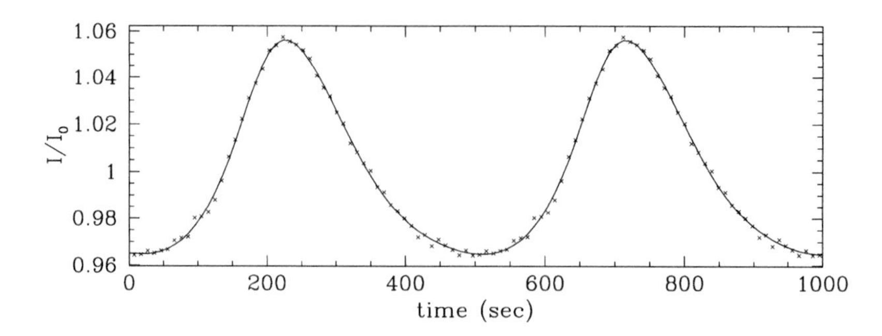

Figure 3: A light curve fit to data taken on PG 1351+489 in May 2004 with the McDonald 2.1m telescope. The data points are shown as crosses and the fit is the solid curve.

Table 2: Results of 1-mode Fit for PG 1351+489.

Epoch	Period (s)	ℓ	m	Amplitude	τ_0 (s)	N	θ_{i}
2004	489.34	1	0	0.30989	89.2	16.0	58.0°
1995	489.34	1	0	0.36529	86.0	20.5	56.7°

The mode periods and identifications which we assumed are given in Table 1, together with the amplitudes derived from the fit. In addition, this fit assumes the observations were made using the SDSS g′ filter, and that $T_{\mathrm{eff}} = 24900$ K and $\log g = 7.91$ for GD 358, which are the best fit zero hydrogen values of Beauchamp et al. (1999). From the calculations in the previous section, this leads us to assume a value of $\alpha_{g'} = 0.259$.

PG 1351+489

Using the updated version of our code, we re-analyzed data taken on the nearly mono-periodic DBV PG 1351+489. The results obtained are very similar to those found earlier (Montgomery, 2005). In Figure 3 we present the fit to the folded pulse shape of the May 2004 data of the dominant 489.34 s peak. In Table 2 we give the values of the fit parameters using this data set and the data set obtained with the WET in 1995. With the exception of the amplitude, we find virtually the same values for the other parameters.

Goodness of Fit and Errors

The fits shown in Figures 2 and 3 "look good", although this is a completely subjective, non-quantitative statement. We would obviously like to know whether these fits really have something to do with the star, and, if they do, what the uncertainties on the measured quantities τ_0, N, and $\theta_{\rm i}$ are, for example.

For this purpose, it is tempting to apply a reduced-χ^2 test to these data. For the NOT data on GD 358, we estimate a point-to-point scatter of about 5 mma, which leads to a reduced $\chi^2 \sim 10$ for the light curve fit. For the folded light curve of the McDonald 2.1m data on PG 1351+489, we estimate a scatter of 1.5 mma, which leads to a reduced $\chi^2 \sim 0.9$. This confirms our feeling that the PG 1351+489 fit is indeed much better than that to GD 358. In fact, the formal probability of a reduced $\chi^2 \sim 10$ is vanishingly small. Should we therefore conclude that our fit is so poor that we do not learn anything about the star, or should we take it to mean that there are simply *additional* effects going on which we have not yet taken into account?

We proceed, naturally, by assuming that our fits *are* telling us something about the star. In an attempt to justify this belief we examine a different diagnostic: we plot how the average squared residuals decrease as fit parameters are added. First we add parameters corresponding to (linear) sinusoidal terms. Each sinusoid has an amplitude, frequency, and phase, so each adds 3 additional parameters. This is shown in Figure 4, in which the decrease in residuals out to 36 parameters is due to the addition of 12 sinusoids to the light curve fits. The right-most point in this plot corresponds to the addition of the (nonlinear) convective fit, which uses the 3 parameters τ_0, N, and $\theta_{\rm i}$. We see that it decreases the residuals by a much larger amount than the previous terms do, so in some sense it encapsulates "more physics". This helps make the case that our fits really do capture some of the essential behavior of the star.

Error estimates are trickier, since they usually involve the assumption that the difference between the data and the fit is due to random, statistical fluctuations in the data. A glance at Figure 2 shows otherwise, however, since excursions above and below the fit tend to contain many consecutive points, not simply one or two as would be expected if the excursions were uncorrelated. This could be due to many things. First, we know that GD 358 contains other, smaller amplitude modes which we have not modelled, and these will cause smooth, coherent departures from the fit. Also, the flux correction factor, $\alpha_{g'}$ (shown in Figure 1), is not a constant as we have assumed, and this will also cause such coherent departures. Finally, there are all the other things which the star is doing which we do not yet know about, and these are undoubtedly also having an effect.

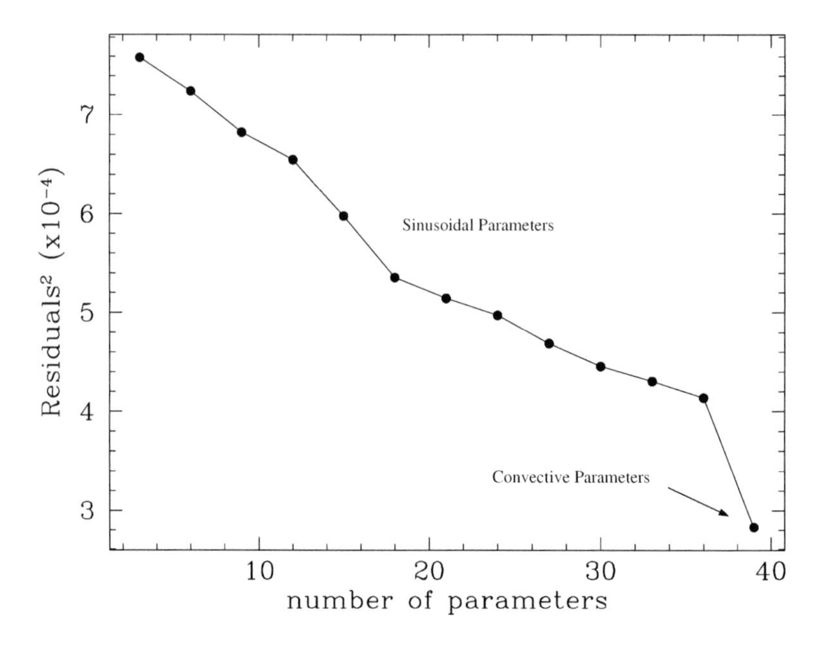

Figure 4: The squared residuals of fits to the light curve of GD 358 as a function of the number of fitted parameters. Each point corresponds to 3 parameters, and the first 12 points are the 36 parameters for the fits of 12 linear sinusoids. The last point shows the squared residuals when the convective effects are added, using the additional parameters τ_0, N, and θ_i. The sharp downturn implies that the light curve fits *do* recover a significant amount of the physics in these objects, more than is obtained from the continuing addition of new frequencies to the fits.

Our current plan is to include the additional effects which we know about, i.e., the variable flux correction and the most important of the lower amplitude modes, and calculate the residuals of the fit. We will then generate synthetic, best-fit light curves to which we have added *uncorrelated*, Gaussian-distributed noise having the same mean value as the residuals. We will then fit this light curve and derive the convective parameters. Finally, we will repeat this process many times to derive error estimates on all quantities; this is the same procedure Montgomery (2005) used to estimate errors for the fits to G 29-38 and PG 1351+489. This work is currently in progress, so we defer the presentation of these results to an upcoming publication.

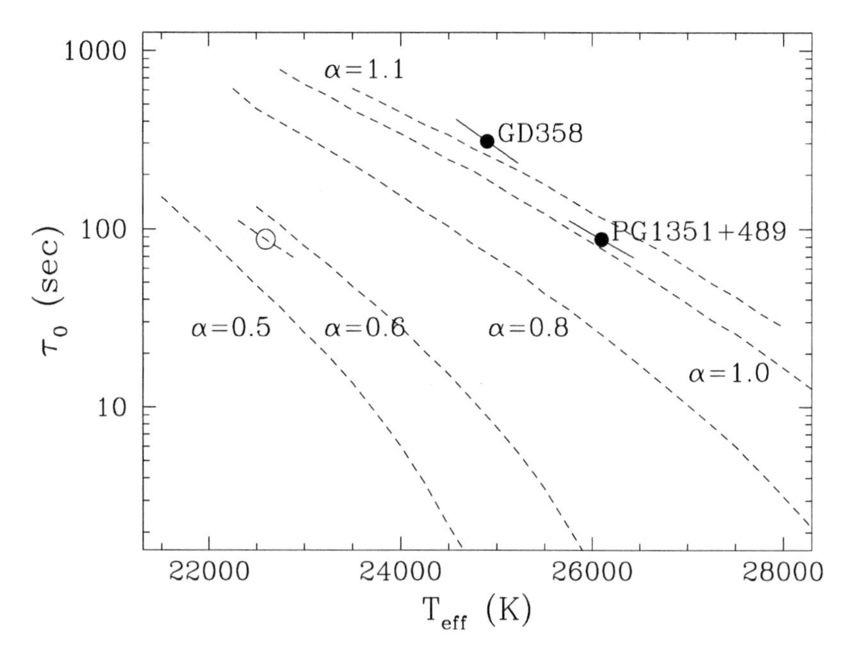

Figure 5: The convective timescale τ_0 as a function of $T_{\rm eff}$ across the DBV instability strip. The labelled points are from the preceding convective fits and the dashed curves are the predictions of standard ML2 convection (Böhm & Cassinelli 1971) and are labelled by the assumed values of the mixing length ratio, α.

Discussion

Now that we have analyzed two DBVs having different values of $T_{\rm eff}$ we can start to see how the convective response timescale changes across the instability strip. The very large caveat here is that in order to do this we need to use published values of $T_{\rm eff}$ and, to a lesser extent, $\log g$ for these stars. Even so, we can make progress.

First, we note that GD 358 has a larger value of τ_0 than PG 1351+489, so it must have a thicker convection zone and therefore be cooler. Due to the possible presence of unseen hydrogen, the spectroscopic $T_{\rm eff}$ determination for PG 1351+489 can be anywhere from 22600 K (with hydrogen) to 26100 K (without hydrogen; Beauchamp 1999). We indicate the position of the cooler solution with the open circle in Figure 5. Clearly, only the hotter solution for this star makes sense. Furthermore, given the slopes (values of N) determined for these stars, we see that the hottest possible solution for PG 1351+489 is preferred, i.e., that with no hydrogen. This demonstrates the power of this analysis in terms of constraining $T_{\rm eff}$.

Second, we see that the locations of these stars are broadly consistent, i.e., we can nearly draw a line connecting the two points having the appropriate slopes. Finally, we can compare these locations with the predictions of simple mixing length theory. As shown in Figure 5, we find ML2/$\alpha = 1.1$ (Böhm & Cassinelli 1971) to be in reasonable agreement with these results.

Acknowledgments. This research was supported by the National Science Foundation under grant AST-0507639 and by a grant from the Delaware Asteroseismic Research Center.

References

Beauchamp, A., Wesemael, F., Bergeron,P., et al. 1999, ApJ, 516, 887

Bergeron, P., Wesemael, F., & Beauchamp, A. 1995, PASP, 107, 1047

Böhm, K. H., & Cassinelli, J. 1971, A&A, 12, 21

Bradley, P. A., & Winget, D. E. 1994, ApJ, 430, 850

Brickhill, A. J. 1992, MNRAS, 259, 519

Holberg, J. B., & Bergeron, P. 2006, AJ, 132, 1221

Metcalfe, T. S., Nather, R. E., & Winget, D. E. 2000, ApJ, 545, 974

Montgomery, M. H. 2005, ApJ, 633, 1142

Montgomery, M. H. 2007a, in Unsolved Problems in Stellar Physics: A Conference in Honor of Douglas Gough, ed. R. J. Stancliffe, G. Houdek, R. G. Martin, & C. A. Tout, vol. 948. AIP, 99

Montgomery, M. H. 2007b, in ASP Conference Series, ed. A. Napiwotzki & M. R. Burleigh, 372, 635

Winget, D. E. 1982, PhD thesis, University of Rochester

Winget, D. E., Nather, R. E., Clemens, J. C., et al. 1994, ApJ, 430, 839

Winget, D. E., van Horn, H. M., Tassoul, M., et al. 1982, ApJ, 252, L65

Wu, Y. 2001, MNRAS, 323, 248

Yeates, C. M., Clemens, J. C., Thompson, S. E., & Mullally, F. 2005, ApJ, 635, 1239

S. O. Kepler and M. H. Montgomery in Mount Cuba Observatory.

Comm. in Asteroseismology
Vol. 154, 2008

On Coordinating Time Series Spectroscopy with the WET

S. E. Thompson[1]

[1]Delaware Asteroseismic Research Center, Mt. Cuba Observatory and University of Delaware, Dept. of Physics and Astronomy, Newark, DE 19716, USA

Abstract

Recently time series spectroscopy of hydrogen atmosphere white dwarfs (DAVs) has proven successful in measuring the spherical degree and pulsation velocity of each mode. The information gained from these observations would be enhanced by having the observations coincide with a multi-site photometry, such as that provided by a Whole Earth Telescope (WET). The accurate photometry would provide the precise periods and photometric amplitudes that cannot be easily obtained by spectroscopy alone. Combining spectroscopy with photometry of DAVs will lead to answers regarding combination modes, convective driving, and the pulsation geometry of each mode. Here I discuss how time series spectroscopy will improve the results of the photometric runs and the quality and quantity needed for success.

Introduction

The Whole Earth Telescope (WET) is an ideal tool for accurately identifying and measuring the pulsation periods of variable stars. A common target of these multi-site campaigns has been hydrogen atmosphere white dwarf pulsators (DAVs). The scientific goals for studying DAVs have included modeling white dwarf (WD) interiors (e.g. Clemens 1993), measuring convection from light curve shapes (Montgomery 2005, 2007), and measuring cooling rates from period changes (e.g. Kepler et al. 1991). The ultimate success of these projects depend, in part, on knowing the spherical degree, ℓ. However since the photometry of the WET does not directly measure ℓ, we must either infer the values from asteroseismic models, or rely on other methods.

Time series spectroscopy (TSS) of WDs has proven successful for measuring ℓ and pulsation velocity. The technique depends on subtle variations in Hydrogen line to distinguish between different values of ℓ and small shifts in the spectral line to measure the velocity. The required high signal-to-noise and short exposure times limit the observations to some of the largest operating telescopes. As such, obtaining long, continuous spectroscopic sets is almost impossible, resulting in poorly measured periods and no identification of closely spaced modes. However, this short-coming can be surmounted by taking the spectra during a WET run and using the multi-site photometry to identify the pulsation periods.

Measurements with TSS

By taking TSS of DAVs we can measure aspects of pulsation that time series photometry cannot. First, Doppler shifts of the spectral lines measure the longitudinal motion of the pulsation, perpendicular to our line of sight. This motion causes the observed brightness variations and thus is a more fundamental measure of the pulsation amplitude. Observing velocities allows us to distinguish those flux variations associated with physical pulsation from those created by interacting with the convection zone (van Kerkwijk et al. 2000). Second, the Hydrogen lines change shape along with the stellar flux variations in accordance to the distribution of the pulsation across the stellar surface. Since limb darkening varies across the spectral lines, each wavelength samples a different area of stellar surface. Each spherical degree, with their unique latitudinal distribution, shows a different amount of cancellation when integrated over the observed surface area. Since the total surface area varies with limb darkening, the amplitude of the pulsation at each wavelength will be unique to that value of ℓ (Robinson et al. 1995).

The most successful TSS white dwarf target is G29-38. As the brightest large amplitude DAV, observations from Keck and the VLT have identified ℓ for 12 different modes from the Hydrogen line shapes (Clemens et al. 2000, Kotak et al. 2002, Thompson et al. 2006). Additionally, van Kerkwijk et al. (2000) and Thompson et al. (2003) have measured the velocity associated with the largest pulation modes. They show that the ratio of the velocity and the flux amplitudes for periods near 800 s are consistent with those predicted by the convective driving theory (Wu & Goldreich 1999).

Part of the success of TSS is exposing new mysteries. Though TSS for mode identification has only been applied successfully to $\approx$16 modes on four stars, 5 of those modes are not ℓ=1, leaving the impression that higher ℓ modes are more common than previously thought. The ratio of the velocity to flux amplitudes have not been accurately measured for shorter periods, but those measurements that do exist indicate that the measurements may not follow the predictions of the convective driving theory (see Thompson et al. 2003). Pulsation velocity measurements of G29-38 have shown that the physical pulsations produce combination velocity modes. Are these velocity combinations unique to G29-38? Are they actually resonant modes, or is there a force causing nonlinear mixing of the physical pulsations of the star?

The results for G29-38 come from runs less than 6 hours, leaving the precision of the measured frequencies to at best 45 μHz. Large amplitude pulsators, like G29-38, tend to show a complex set of pulsation modes that vary between observing runs. To know the true pulsation structure of the star, we need longer coverage, but we do not necessarily need more observations if a WET run coincides with the TSS observations. The WET can provide the extended coverage to measure the frequencies of pulsation. If TSS is taken over several nights (e.g. 3 nights provides a frequency resolution of 6 μHz), the worry of which alias peak to select among the spectroscopic FTs is not a problem, the WET would have already identified and measured the frequencies. In this way extended photometry will improve the output of the TSS while the TSS will provide crucial information about ℓ and pulsation physics that ultimately will help decipher the star.

Among the WD pulsators, the DAVs, with their deep hydrogen lines, show the greatest variation between different spherical degree modes. Also, with the abundance of known DAVs (see Castanheira et al. 2006, Mullally et al. 2005, and Mukadam et al. 2004), we are more likely to measure spherical degrees and pulsation velocities on an interesting number of targets.

Measuring spherical degree

Thompson et al. (2004) introduced a new technique that obtains ℓ from data that previously had been of too poor of quality to perform the task. By fitting the spectra with the combination of a Lorentzian and a Gaussian, they focused on only the necessary information contained in the spectra to determine the spherical degree. In this case they simplified the periodic line shape variations by only allowing the height of the continuum along with the areas of the Gaussian and the Lorentzian to change from the initial fit to the average spectrum. When compared to models, it is clear that the measurements do not accurately agree with the observations. However, for G29-38 most modes have the same line variations, and most closely resemble the ℓ=1 model (Clemens et al. 2000).

As a simple method to demonstrate the relationship between the observations and the models, we measure the Gaussian equivalent width (EW) and the Lorentzian EW from the spectral fits. We then plot the amplitudes of these variations against each other after normalizing by the flux amplitude of

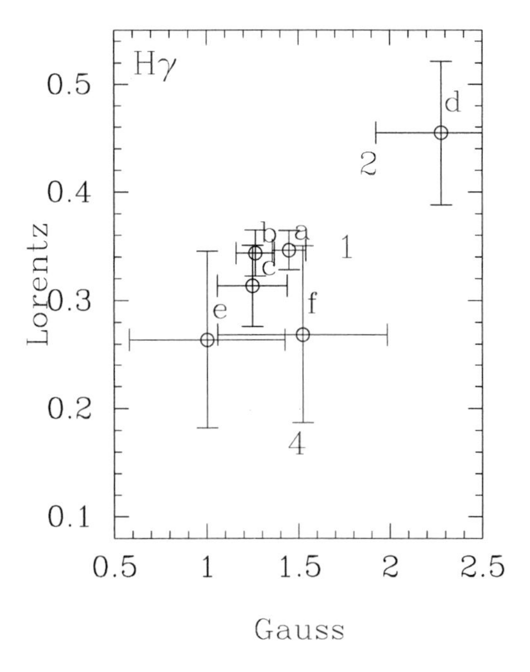

Figure 1: The normalized Gaussian EW vs. the normalized Lorentzian EW amplitudes measured for the Hydrogen-gamma line for pulsated model DAV spectra (numbers), and for observations of G29-38 observed with the Keck telescope (a-f) (Clemens et al. 2000). The different ℓ as established with pulsated atmospheric models of DAVs are labelled according to their value. Only the six largest modes were measured in this manner. Most modes cluster near the ℓ=1 mode, though show a distinct offset from the model. Mode 'a' (614 s) is ℓ=1 while mode 'd' (776 s) is ℓ=2.

the mode. Fitting pulsated model spectra of a DAV shows that these amplitudes are different for different ℓ. See Figure 1 for a comparison of the model to observations of G29-38. The majority of modes are ℓ=1 while the one ℓ=2 mode has larger Gaussian and Lorentzian EW amplitudes (Clemens et al. 2000 and Thompson et al. 2006). The models agree with this trend, showing ℓ=2 with larger Gaussian and Lorentzian EW amplitudes. The ℓ=1 identification of the 615 s mode is supported by the convective fitting analysis of Montgomery (2005). The ℓ=2 measuremnt for the 776 s mode is supported by the especially large velocity amplitude for this mode (van Kerkwijk et al. 2000).

With this analysis we have a simple way to determine the quality of data needed to distinctly measure ℓ. The quality of data affects our ability to accurately fit each spectrum and thus the size of our error bars. As long as our error bars are smaller than the distance between the ℓ=1 and ℓ=2 models, we are likely to make the correct measurement of ℓ. At this modest resolution, most spectrographs will provide a wide enough wavelength range to allow us to measure more than one spectral line, thus we can obtain 2-3 measurements of ℓ for each mode from one set of TSS.

TSS Quantity and Quality

By running simmulations of noisy model spectra, I determine the quantity and quality of the spectroscopy needed to measure ℓ with TSS. Aspects of the measurement that affect the ability to measure ℓ include: the signal-to-noise of the individual spectrum, the resolution of the spectrum, the total number of measurements, and the time span these measurements cover. For our example here, let us assume the time span is enough to accurately distinguish all frequencies, and are known accurately from supporting photometry. From previous experience we find that only a modest resolution (≤ 7 Å) is necessary to measure the changes in line shape. This resolution is usually set by the seeing since a wide slit is used to gather as much light as possible. Getting the required S/N is the tricky part; to reach more than ten DAVs requires observing stars as dim as 16 mag. Eight meter class telescopes can obtain a S/N of 70 for a 15.5 mag star with a 30 s exposure; the cycle time is then near 50 s.

I ran simmulated spectra with a resolution of 6 Å (and a dispersion of 1.8 Å/pixel), S/N of 70, and cycle time of 50 s to determine the total length of observations necessary to measure various amplitude pulsation modes on a DAV. The expected significance of our ℓ measurements can be found in Figure 2. Since the goal is to uniquely measure ℓ, we judge significance based on whether an ℓ=2 could be distinguished from an ℓ=1 in the Hγ line. However, remember it is likely that the same spectra will also measure Hβ and Hδ. The ordinate in Figure 2, labelled "significance" is the ratio between the measured error bar and the difference between the ℓ=1 and ℓ=2 models for the Hγ line. To observe the 8 mma (0.8 %) modes, we would need almost 16 hours of telescope time. However with only 4 hours, we could measure the spherical degrees of the large 20 mma modes.

Measure Velocities

The same spectra that measures ℓ can be used to measure the velocity of each mode. However, this is not the best way to measure velocities. While sacrificing resolution for more light improves our ℓ measurements, it deteriorates the velocity measurements. Thompson et al. (2003) showed that with higher resolution spectra (0.215 Å, S/N$\sim$8, 5 hrs of observations), they could significantly improve the velocity measurements. For the higher resolution data, they

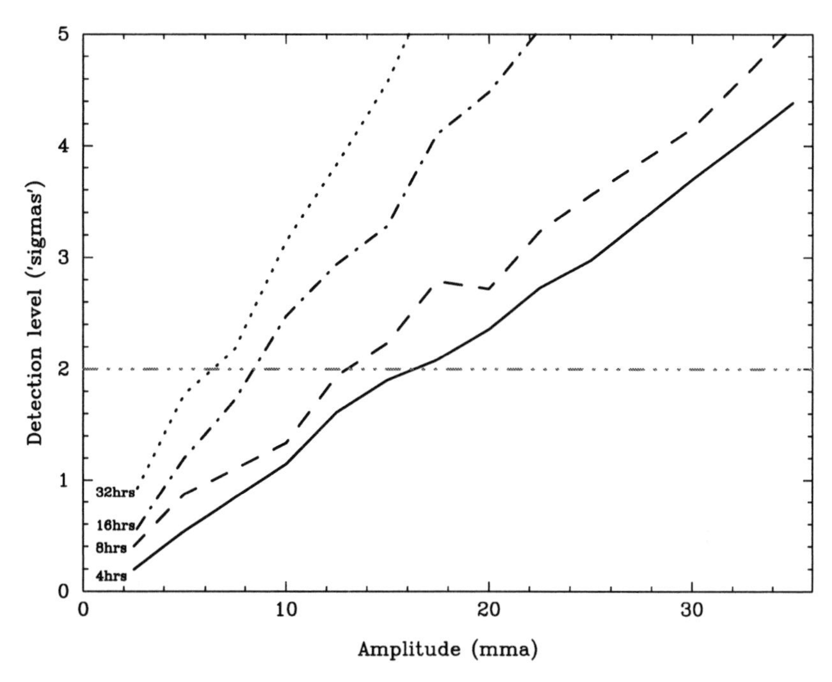

Figure 2: The TSS detection significance to distinguish between ℓ=1 and ℓ=2 of different amplitude modes given various length of observing runs. These values come from simulations of noisy Hγ spectra with a S/N of 70 and an observational cycle time of 50 s. The simmulation assumes the pulsation periods are known either from the spectroscopy or coincidental photometry.

could detect velocity amplitudes down to 1 kms^{-1} while for the lower resolution spectra (~7 Å, van Kerkwijk et al. 2000) they were limited to those above 2 kms^{-1}. The lower noise level revealed velocities for modes with flux amplitudes as small as 10 mma. To obtain this higher resolution you must observe with a slit and reduce the spectral range, both significantly reduce the amount of light gathered, and deteriorate the measured flux amplitudes.

By measuring the velocities, we can explore the nature of the combination modes and the validity of the convective driving theory proposed for DAVs. The velocity, or lack thereof, will reveal which modes are physical pulsations on the star and which are combination modes. Combinations generally have shorter periods and lower amplitudes, requring higher quality velocity measurements. The ratio between velocity amplitudes and flux amplitudes seem to agree with observations for the longer periods (~800 s) but have not been adequately tested for the short periods. Unfortunately these short period modes tend to have smaller flux and velocity amplitudes. The limiting factor from the previous high resolution spectra was the poor light curve. At least for G29-38, velocity measurements of a significant number of short period, lower amplitude modes is possible, but requires coordination with a set of photometric observations to obtain the flux amplitudes and precise frequencies.

Conclusions

TSS, though a fairly new tool for DAVs, has proven its effectivness, especially for the DAV G29-38 by measuring velocity pulsations and ℓ. Both high and low resolution TSS have measured the pulsation velocities of G29-38. In the future, if such studies are combined with photometry, they could provide conclusive tests of the convective driving theory, and further examination of the nature of the combination modes.

The improvements in TSS by fitting spectra and spectrograph efficiency make it possible to perform ℓ identification for currently about ten DAVs. Mode identification is generally inferred from the very models that are used to constrain the stellar structure. While blindly choosing ℓ=1 is a good guess, now that we know of at least three ℓ=2 and two $\ell \geq 3$ modes (Thompson et al. 2004; Kotak et al. 2002; Thompson et al. 2008) after successful TSS analysis of only 16 modes on DAVs, we must question this guess. If the goal of the WET, or any multi-site photometric study, is to perform asteroseismology of a DAV, I encourage the study to pursue TSS to improve the validity of their mode identifications and ultimately their measurements of stellar structure.

TSS must rely on additional photometry to accurately know the pulsation frequencies, however photometry must rely on spectroscopy to measure the physical pulsation velocity and ℓ. The coordination of these two observations will improve the information gained from either acting alone.

Acknowledgments. I acknowledge M. H. van Kerkwijk for proving the reduced Keck spectra of G29-38 and the Crystal Trust for their contribution to the completion of this work.

References

Castanheira, B. G., Kepler, S. O., Mullally, F., et al. 2006, A&A, 450, 227

Clemens, J. C. 1993, Baltic Astronomy, 2, 407

Clemens, J. C., van Kerkwijk, M. H., & Wu, Y. 2000, MNRAS, 314, 220

Goldreich, P., & Wu, Y. 1999, ApJ, 511, 904

Kepler, S. O., Winget, D. E., Nather, R. E., et al. 1991, ApJL, 378, L45

Kotak, R., van Kerkwijk, M. H., & Clemens, J. C. 2002, A&A, 388, 219

Montgomery, M. H. 2005, ApJ, 633, 1142

Montgomery, M. H. 2007, Astronomical Society of the Pacific Conference Series, 372, 635

Mukadam, A., Mullally, F., Nather, R. E., et al. 2004, ApJ, 607, 982

Mullally, F., Thompson, S. E., Castanheira, B. G., et al. 2005, ApJ, 625, 966

Robinson, E. L., Mailloux, T. M., Zhang, E., et al. 1995, ApJ, 438, 908

Thompson, S. E.,Clemens, J. C., van Kerkwijk, M. H., & Koester, D. 2003, ApJ, 589, 921

Thompson, S. E., Clemens, J. C., van Kerkwijk, M. H., et al. 2004, ApJ, 610, 1001

Thompson, S. E, van Kerkwijk, M. H., & Clemens, J. C. 2008 MNRAS, submitted

van Kerkwijk, M. H., Clemens, J. C., & Wu, Y. 2000, MNRAS, 314, 209

Wu, Y., & Goldreich, P. 1999, ApJ, 519, 783

F. Mullally and S. E. Thompson taking notes in the back of the room.

Comm. in Asteroseismology
Vol. 154, 2008

Preliminary Results in White Dwarf Research by a New Group

M. Paparó[1], E. Plachy[1,2], L. Molnár[1,2], P.I. Pápics[2], Zs. Bognár[1], N. Sztankó[2], Gy. Kerekes[2], A. Már[2], and E. Bokor[2]

[1] Konkoly Observatory, Budapest, P.O. Box 67, H-1525, Hungary
[2] Eötvös Loránd University, Budapest, Pázmány Péter sétány 1/A, H-1117, Budapest, Hungary

Abstract

A new group, named PiStA, has been established for investigation of white dwarf stars by undergraduate students. Beside a contribution to WET/DARC campaigns, an independent research program has been started. Up to now data collection has been carried out for 5 stars (GD 154, GD 244, KUV 02464+3239, LP 133-144 and G207-9) over a whole observational season for each target.Preliminary results are presented for GD 154. An additional frequency at 213.6 c/d has been identified in an early data set. An alternative interpretation on all available data comparing with a pure multiperiodic pulsation is in progress. Sign of a chaotic behaviour is going to be investigated.

Introduction

Pulsating variable star research has a long tradition in Konkoly Observatory dating back to the early observation of RR Lyraes and δ Cepheids in 1929. Changes in the instrumentation introduced newer and newer fields in the Observatory's profile. As theoretical interpretation of non-radial pulsation developed and the precision of the photoelectric photometry increased, the investigation of low-amplitude, short period, multiperiodic δ Scuti stars, pulsating mostly in non-radial modes, has been started beside the large amplitude, radially pulsating RR Lyraes, δ Cepheids and high amplitude δ Scuti stars (HADS). Paparó joined many international campaigns in the framework of Delta Scuti Network. However, the instrumentation changed again. The 1 m RCC telescope equipped with the CCD camera allowed us the investigation of faint, short period, low amplitude white dwarfs. As a first activity some of the stuff members joined WET campaigns (Handler et al. 2003, Fu et al. 2007). Single site data collection was also urged by some local scientific political reasons. WET/DARC international campaigns have superior results on white dwarfs due to two or three-week long continuous datasets with excellent coverage.

Figure 1: Designed by the team. Light curve of GD 154 is used from our own observation.

The enthusiasm of undergraduate students of Eötvös University on the local, single site observations resulted the establishment of Piszkéstető Students' Astronomy (PiStA) group as a student scientific project (Figure 1). The present members are on masters level. A new generation of PISTA, on even bachelor level, is going to join the group these days. Of course, it takes time to train the students to reach the scientific level and it can be done only step by step as the data processing is going on. Up to now our team (my PhD student, Zsófia Bognár and PiStA) has collected remarkable amount of data for many white dwarfs. We have started the final work on GD 154.

Single-site observation of white dwarf stars

White dwarf research is dominated by a worldwide organization, WET/DARC. When applying for telescope time we have to argue why we need the telescope for concentrated and coordinated observation of a single star. The most important argument is to avoid aliases that are caused by gaps in the data distribution since detection and identification of real pulsation frequencies are much more difficult if aliases are present.

However, there is another fact to be considered. A long light curve is necessary to resolve the closely spaced frequencies in multiperiodic stars. WET/DARC campaigns last no longer than 2-3 weeks. A possible solution is to have two campaigns separated by a short interval (two months for PG 0122+200) to improve the frequency resolution and to get higher precision on the frequencies (Fu et al. 2007). Nevertheless, it is not easy to organize telescopes (especially larger ones) too often to work on the same target. Space projects would be also a solution. According to our knowledge neither the running nor forthcoming space projects will observe white dwarf stars.

Single-site monitoring of white dwarfs is still worthwhile. We can follow their pulsational behaviour not in a continuous dataset but on a long time base. We can have a sample of temporal characteristics from month to month up to 5-6 month if the target is properly chosen in declination. We can always use the best part of the sky. Of course, we are faced with the problem of side-lobes and with a beating of closely spaced frequencies over shorter datasets. However, careful comparative analyses can help to come over part of the problem.

It has become clear in the last years that amplitude and frequency variations are common amongst pulsating stars. White dwarfs are not exceptional in this respect either. Although, some of the modes are extremely stable and the period increase and decrease can be measured, the amplitude variability and in some cases the frequency variability are well-established (down to two weeks, Handler et al. 2003). The aim of our "continuous" single-site observational project is to follow the short-term variability of pulsation in white dwarfs. Of course, we will join the WET/DARC campaigns from time to time.

Observed targets

We used different guidelines in target selection. The main aim was to keep up the enthusiasm of the young group. GD 154, the first target, was suggested by Gilles Fontaine for educational purposes. The star has short periods and large amplitudes. The aim was fulfilled. The students were delighted from the first moment by the impressive light variation of GD 154. We later followed more practical and more scentifical guidelines as criteria. We always wanted to use the best part of our sky. Regarding the faintness of white dwarfs and the size of our telescope, we had to pay attention to this point. We usually picked the newly discovered stars with coordinates proper to our site.

Table 1 gives the number of clear nights over the given interval. Each target was observed in the whole observational season, as often as we got telescope time at our mountain station at Piszkéstető. The observations were carried out by a Roper Scientific 1300B CCD camera (20*20 μm, 1340x1300 pixel) with a 7'x7' field of view. In each cases we observed in white light to have as many counts as possible. Ten or 30 seconds exposure time, depending on the quality of the sky, were used. In average, we used approximately 50% of the telescope time allocated to our targets, which is not a bad ratio.

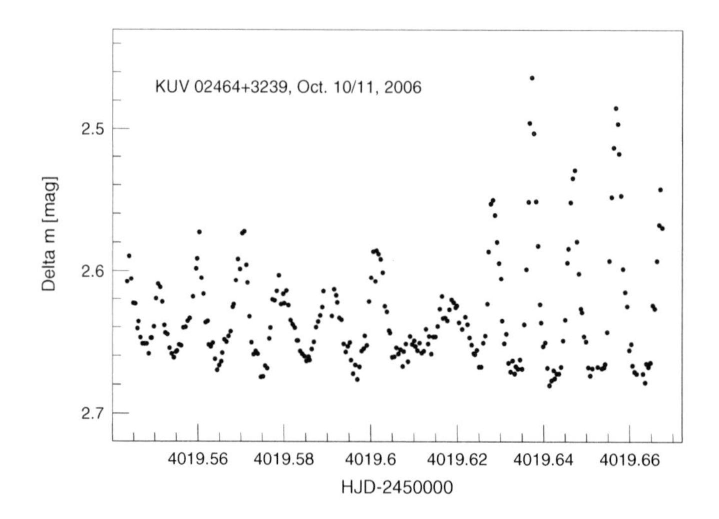

Figure 2: KUV 02464+3239 has large amplitude, long period. It is located near the red edge of DAV instability strip.

Table 1: Targets observed in 2006-2007 by PiStA.

Star	**No. of nights**	**Interval**
GD 154	20	Febr. - July 2006
GD 244	14	July - Oct. 2006
KUV 02464+3239	19	Oct. 2006 - Febr. 2007
LP 133-144	28	Jan. - May 2007
G207-9	24	March - Aug. 2007

A sample of the light curves obtained for our targets are presented. A typical part of the light curve of KUV 02464+3239 is given in Figure 2. It has a large amplitude and a long period in agreement with its location near the red edge of the DAV empirical instability strip. A preliminary analysis of part of the presently available data was presented by Zs. Bognár et al. (2007). Even in the preliminary analyses of a short dataset three additional frequencies were obtained. The final analysis of KUV 02464+3239 is in progress.

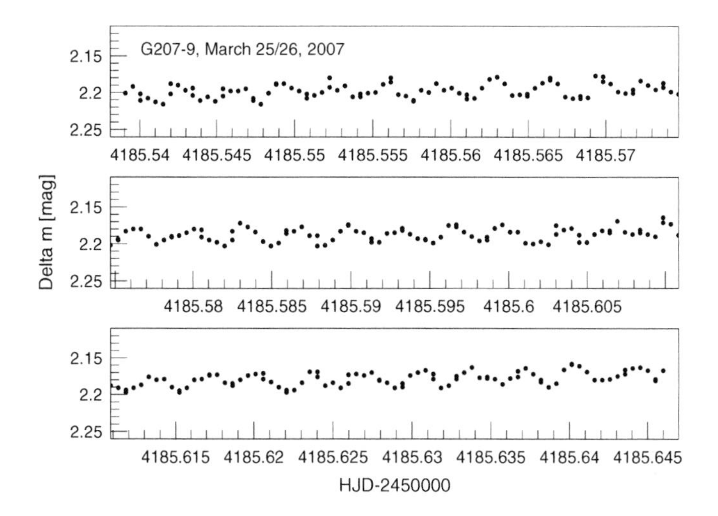

Figure 3: High presision measurements allowed us to follow the low amplitude and short period of G207-9.

Figure 3 shows a sample of the light curve of the DA white dwarf, G207-9. It has much lower amplitude than KUV 02464+3239. Our precision is still good enough to follow the short period and low amplitude single cycles.

LP 133-144 has been recently discovered (Bergeron & Fontaine 2004). The discovery paper shows a rather irregular light curve just as our observations (Figure 4). As they stated, the relatively short period and low amplitude are consistent with the star's location near the blue edge of the ZZ Ceti instability strip.

According to the first detection, GD 244 is an intermediate-amplitude pulsator with periods which are consistent with its position in the middle of the ZZ Ceti instability strip (Figure 5). It appears to be a photometric clone of GD 66. As a result of this WET/DARC workshop, a collaboration was established. F. Mullally allowed us to work on his unpublished data of GD 244 (2003 - 2006). Our team obtained data for GD 66 in a multisite, coordinated run (October-November 2007).

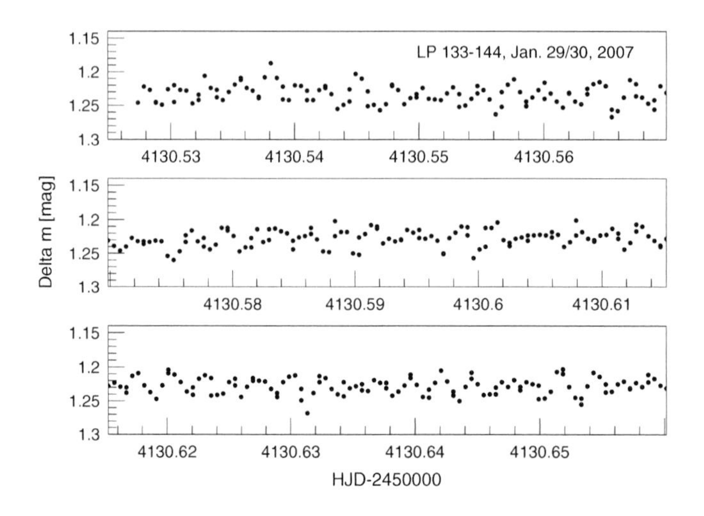

Figure 4: LP 133-144 has an irregular light curve like in the discovery paper. It is near the blue edge of DAV instability strip.

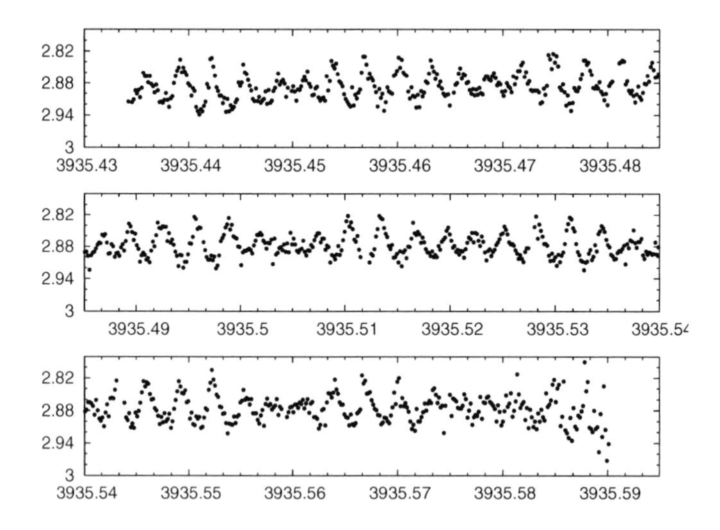

Figure 5: GD 244 is an intermediate-amplitude pulsator located in the middle of the DAV instability strip.

Preliminary results on GD 154

GD 154 belongs to one of the best studied DAV white dwarfs ($T_{eff} = 11180/$,K, log g = 8.15, M_v=15.33, M=0.7$M_{\odot}$). It was discovered in 1977 (Robinson et al. 1978). Observations were obtained on ten nights (23.5 hours) over two months. Probably the abrupt change in the rather regular light variation made the star interesting enough to have it as a WET target in 1991 (Pfeiffer et al. 1996). In the WET campaign 162.5 hours data was collected in two weeks from 6 observatories. A shorter run (62 hours over 12 nights) was carried out in 2004 from two sites (Hürkal et al. 2005).

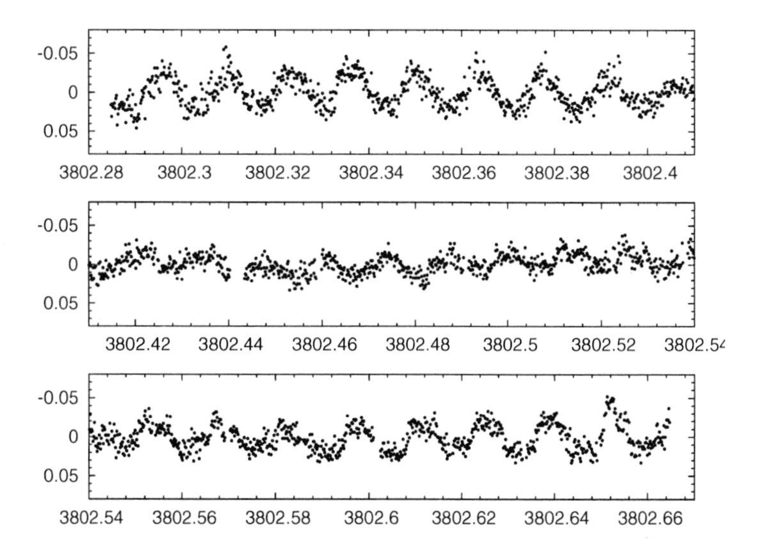

Figure 6: Light curve of GD 154 with 10 s exposure time.

Observation of GD 154 was carried out at Piszkéstető, the mountain station of Konkoly Observatory, from February to July in 2006. To have an impression of the quality, a single night data is shown in Figure 6. Altogether 90 hours data was collected over 20 nights, using 16315 CCD frames. If we compare with the previously obtained datasets, our dataset is not as long and the distribution is not as compact as the others, but it covers a much longer interval of the star's pulsational behaviour.

In data reduction the IRAF package was used but not in a direct command mode. To avoid any personal mistake and to speed up the process a Perl script was used for the reduction. Final data processing, especially the correction for colour effects, is going to be finished for the end of 2007. Regarding the B-V = 0.19 value of GD 154, we found that the B-V values of the potential comparison stars on the field are rather different (from 0.59 to 1.37). These values are converted from the SDSS colours. Careful colour correction is needed.

We are planning to have a comparative analysis for all available data, however, to have the previous data in final form proves not to be a trivial task. The raw data was kindly put at our disposal by Robinson and Kawaler. New reduction has been finished (data from 1977) or is almost finished (data from 1991), respectively.

To get a homogeneous aspect of the star's pulsational behaviour we started the independent analyses of datasets divided by many years. Obviously we started with the earliest dataset. The dominant feature of the light curve are shown in Figure 7. Most of the time they seem to have a single period. Deviation from a single sine wave suggests the presence of the harmonics, too. However, the most remarkable feature is the alternately lower and higher amplitude value of the consecutive cycles. It is similar to the period doubling bifurcation that happens on the way to the chaotic pulsation. Subharmonics of the frequencies are also expected.

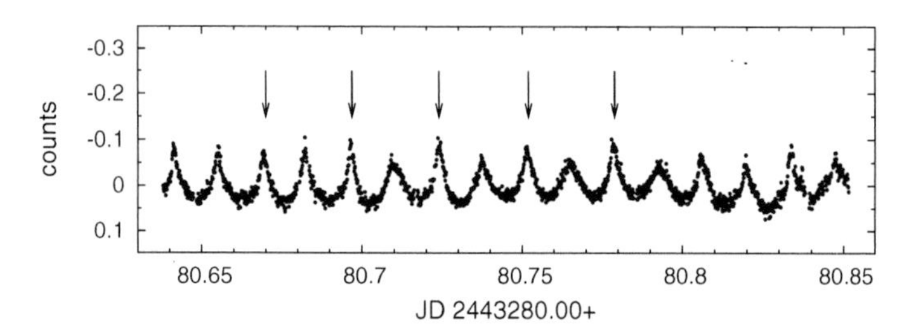

Figure 7: A typical part of the light curve of GD 154 in 1977.

Of course, the result of our Fourier analyses (Table 2) confirmed the previous statements. In agreement with Robinson et al. (1978) we found the dominant frequency at 72.83 c/d. Harmonics up to $6F_1$ are also significant. Subharmonics, not only at $1.52F_1$ but its harmonics at $2.52F_1$, are also above the significance level. The peak at $3.52F_1$ has just a little lower significance value than the significance level. Subharmonic at $1.52F_1$ was interpreted by them as an independent frequency. This view was highly supported by the fact that this was the dominant mode on the last night of their observational run (the light curve had completely changed).

As a new result, we localized a second independent frequency at 213.61 c/d, which is near the $3F_1$ value (218.51 c/d). It is confirmed by the significance value. In the paper of Robinson et al. (1978) the Fourier spectrum of a single night is shown, to present the harmonics and subharmonics. We used the whole dataset in our analyses, except the drastically different last night.

Table 2: Pulsation modes observed in GD 154.

	Frequency	**Amplitude**	**S/N**
F1	72.8381 ± 0.0001	0.0334 ± 0.0002	38.30
2 F1	145.6762 ± 0.0004	0.0114 ± 0.0002	18.87
3 F1	218.5142 ± 0.0007	0.0060 ± 0.0002	11.70
4 F1	291.352 ± 0.001	0.0038 ± 0.0002	8.23
5 F1	364.190 ± 0.002	0.0024 ± 0.0002	6.61
6 F1	437.028 ± 0.003	0.0013 ± 0.0002	4.05
1.52 F1	110.7449 ± 0.0008	0.0052 ± 0.0002	7.40
2.52 F1	183.588 ± 0.002	0.0026 ± 0.0002	4.80
F2	213.610 ± 0.002	0.0025 ± 0.0002	4.91
$\sim$ *F1*	74.211 ± 0.001	0.0038 ± 0.0002	4.36
3.52 F1	256.455 ± 0.002	0.0017 ± 0.0002	3.73

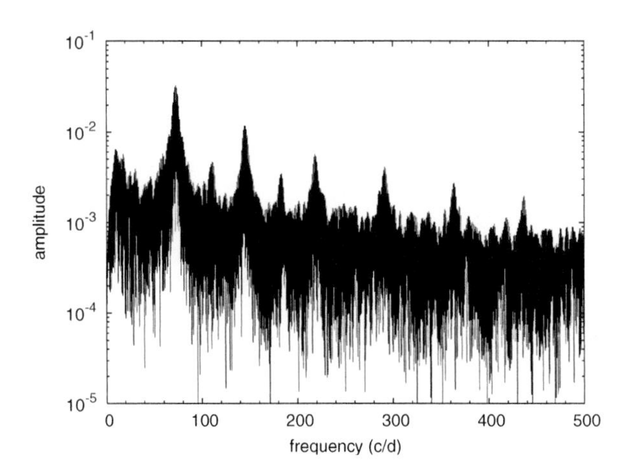

Figure 8: Harmonics and subharmonics are clearly seen.

The second frequency is not a new one, it was found in the WET dataset (Pfeiffer et al. 1996). As the subharmonics appear more clearly on the logarithmic scale, it is shown in Figure 8. The two subharmonics are clearly seen but the $0.5F_1$ subharmonic is missing. Subharmonics were not found in the compact dataset of WET campaign in 1991. Only three independent frequencies and their linear combinations were identified. However, even in the preliminary analyses of our dataset obtained in 2006, we also found a sign of subharmonics.

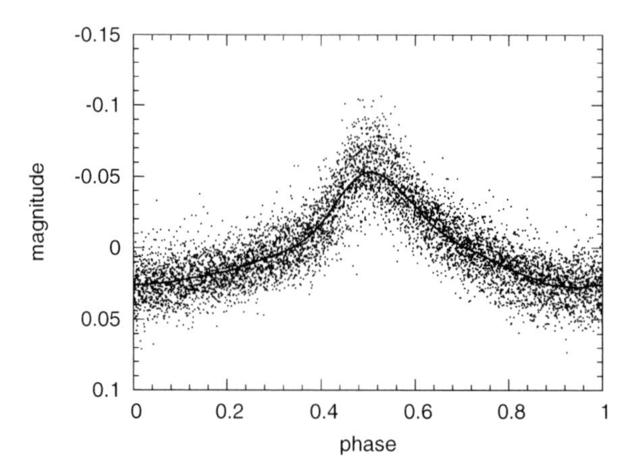

Figure 9: Folded light curve of GD 154.

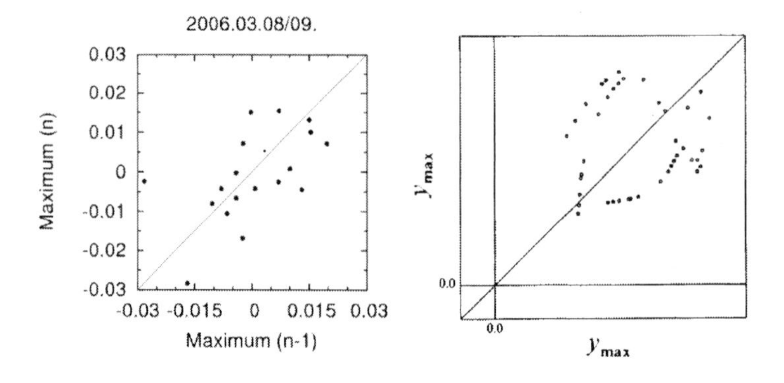

Figure 10: Empirical return map of GD 154 (left) and a theoretical one of a pure chaotic system (right).

As a first step of the nonlinear investigation planned for all available data, the folded light curve of GD 154 using data from 1977 is given in Figure 9. Maybe, it is unusual in the white dwarf's investigation but it is widely used for RR Lyrae stars, to get information on the systematic cycle to cycle variation, like Blazhko effect. It is a basic step to build up the return map, i.e. the plot of the values of the consecutive maxima versus those of the previous maxima. This is the simplest method where the chaotic behaviour can be investigated. Return maps are shown in Figure 10. The left panel shows a return map based on our single night observation. Return map of pure chaotic system by Saitou et al. (1989) is given in the right panel. A similarity of the empirical return maps of GD 154 to a pure chaotic system suggests that it is worthwhile to investigate in details. We will report on the more sophisticated investigation of the chaotic behaviour of our final data in a next paper. Chaotic behaviour has been investigated in white dwarfs (Goupil et al. 1988) but a pure period doubling bifurcation has never been definitely proved in any case. GD 154 is a promising candidate.

Conclusion

The continuous single-site monitoring seems to be worthwhile in characterizing the temporal pulsational behaviour of white dwarfs. It is an excellent scientific project for students providing them with their master or PhD thesis. What is more, they can obtain high level scientific results.

Acknowledgments. We are grateful to Zoltán Csubry for computational support and to Zoltán Kolláth for helpful discussions on the chaotic behaviour of pulsating stars.

References

Bergeron, P., & Fontaine, G. 2004, ApJ 600, 404

Bognár, Zs., Paparó, M., Már, A., et al. 2007, AN, 328, 845

Fu, J. -N., Vauclair, G., Solheim, J.-E., et al. 2007, A&A 467, 237

Goupil, M. J., Auvergne, M., & Baglin, A. 1988, A&A 196, 13

Handler, G., O'Donoghue, D., Müller, M., et al. 2003, MNRAS 340, 1031

Hürkal, D.Ö., Handler, G., Steininger, B.A., et al. 2005, ASPC 334, 577

Pfeiffer, B., Vauclair, G., & Dolez, N. 1996, A&A 314, 182

Robinson, E. L., Stover, R. J., Nather, R. E., et al. 1978, ApJ 220, 614

Saitou, M., Takeuti, M., & Tanaka, Y. 1989, PASJ 41, 297

M. Paparo, S. O. Kepler and G. Vauclair discuss the last talk.

Comm. in Asteroseismology
Vol. 154, 2008

Recent observations of WD1524 at MIRO and participation in WET observing program

H. O. Vats & K. S. Baliyan

Physical Research Laboratory, Ahmedabad 380009, INDIA

Abstract

Most stars end up as white dwarfs on their evolutionary track. A knowledge of the structure of white dwarfs puts constraints on their prior evolutionary stages. By conducting co-ordinated continuous long term observations of these pulsating objects, high precision light curves can be generated to study their environments. We report preliminary photometric observations of WD J1524 from Mt Abu Infrared Observatory (MIRO). The photometric data collected for this source during 27-28 May 2007, though not of sufficient duration, indicate a periodicity of 210 s. These observations provide us confidence to join the future observing programs of Whole Earth Telescope (WET) for such studies.

Introduction

Asteroseismology via precise photometry provides a useful tool to probe the interiors of the pulsating stars (Winget 1998). The white dwarfs (WD) show pulsations, with a range of periodicity frequencies and several tens of millimag amplitude variations which makes them good candidates for the study of stellar evolution. Most of the stars in the universe end up as white dwarfs in their evolutionary phase, and their structures can be used to put constraints on their prior evolutionary stages. As they are faint objects only the nearby objects have been detected so far. However, a large number of white dwarfs should be present in the Galaxy. Determining the contribution of white dwarfs to the total mass of the Galaxy could help solve one of the fundamental questions in modern astronomy: presence and extent of dark matter. The properties of white dwarfs, high gravity and temperature, are important means for the study of physics under extreme conditions. Therefore a sample of white dwarfs are

Figure 1: A view of the Mt. Abu Infrared Observatory (MIRO).

being observed in the Whole Earth Telescope campaign (Mukadam et al. 2006) to get as precise and complete a light curve as possible.

Observations

In an attempt to study the feasibility and exploratory aspects of our joining the WET campaign, optical photometry observations of white dwarf WD1524 were carried out from the Mt Abu Infrared Observatory, Mt Abu, Rajasthan, India.

Mt. Abu Observatory

Mt Abu Infrared Observatory (MIRO) is on top of the Gurushikhar peak (Longitude: 72:46:47 E, Latitude: 24:39:10 N) of Aravali range in Rajasthan state of Western India. It houses a 1.2 m Cassegrain focus f/13 telescope, optimized for infrared and optical observations (Deshpande 1995). The Observatory is situated near the highest peak in Aravali range of mountains at 1680 m height above the mean sea level. A view of the observatory is shown in Figure 1.

The Mt. Abu telescope is an equatorially mounted open truss and fork type and has a 1.2 m parabolic primary (f/3) and 300 mm hyperbolic secondary. It has a Cassegrain focus (f/13) behind the primary at a nominal distance of 380 mm from the Cassegrain plate/instrument ring (Deshpande 1995). At 15.6 m effective focal length of the telescope, the plate scale is 13"/mm that remains linear over the telescope field of view of 10 arcmin diameter. The diameter of the primary central hole is 230 mm. The primary was polished from a mirror-blank of Cervit and weighs 300 kg whereas the secondary made up of Zerodur weighs 6 kg. The primary mirror is mounted in a mirror cell enclosure and is supported on floating axial (18 nos.) and radial (12 nos.) support pads. Each supporting pad is connected to an astatic lever and a balancing counterweight. The mirror support system allows the mirror to float inside the mirror cell with maximum deformation of 20 nm over the entire mirror surface against the cell flexure in any telescope position. The secondary mirror is mounted with a mounting ring at a distance of 2820 mm in front of the primary and is supported by 4 streamlined radial wings. For mounting a back-end instrument on the telescope, an instrument ring of 450 mm diameter is provided. The instrument ring can be rotated about the optical axis of the telescope as it is attached to a bearing, which is fixed in the plate of the cell. The bearing is of a heavy-duty type and can allow an instrument as heavy as 150 kg to be mounted at the back. To accommodate the back-end instruments of different focal lengths, the secondary mirror can be moved up (20 mm) and down (60 mm) remotely from a console by a stepper motor drive giving a total range of 300 to 2200 mm from the Cassegrain plate.

The telescope was commissioned in 1995 and since then regular astronomical observations are being made. Observations are being carried out using different observing techniques viz. photometry, imaging, spectroscopy and polarimetry. The back-end instruments that are currently in operation at the Mt. Abu Observatory on a regular basis are as follows: (i) NICMOS Camera System (256 x 256 pixel IR Array) (ii) Two-channel infrared fast-photometer (iii) Imaging Fabry-Perot Spectrometer (iv) Fiber-Linked Astronomical Grating Spectrograph (FLAGS) (v) Optical Photo-Polarimeter and (vi) CCD Camera System. The CCD Camera is used for the variability observations of AGNs, galaxies, stars clusters etc. For the observations of WD J1524 at MIRO we used this CCD camera (Baliyan, Joshi & Ganesh 2007) details of which are given in the following:

Pixcellent Imaging Ltd, Cambridge EEV CCD 55-30 Grade 0
1296 x 1152 pixels
Pixel size 22.5 microns
Active area 28 x 26 mm
Read noise 4 electrons rms
Pixel scale 0.48 arcsec/pixel
The CCD is back-end illuminated, LN2 cooled

The filter wheel attached to the CCD is automated and has 6 slots for UBVRI filters alongwith an open slot. The field of view is about 6'x5'.5. Usually on-chip 2x2 binning is used for normal operation. The detector is cooled by liquid nitrogen to below -110 degrees to minimize the dark current.

WD J1524-0030

As mentioned earlier, the structure of white dwarfs can be used to put constraints on the prior evolutionary stages of such stars. The detection of pulsation frequencies through precision photometry is used as a tool in this study. Keeping in mind the interesting science such a study of white dwarfs leads to, we at the Mt Abu Observatory also took part in the WET program and during some spare time (about 1.5 hours) WD J1524 was monitored. As these were only preliminary and feasibility-testing observations, these as such do not provide any useful input to the WET program on this source. Nonetheless, the source was observed during May 27-28, 2007 at MIRO in R,V and B filters and corresponding light curves were studied. A typical observed CCD field image is shown in Figure 2.

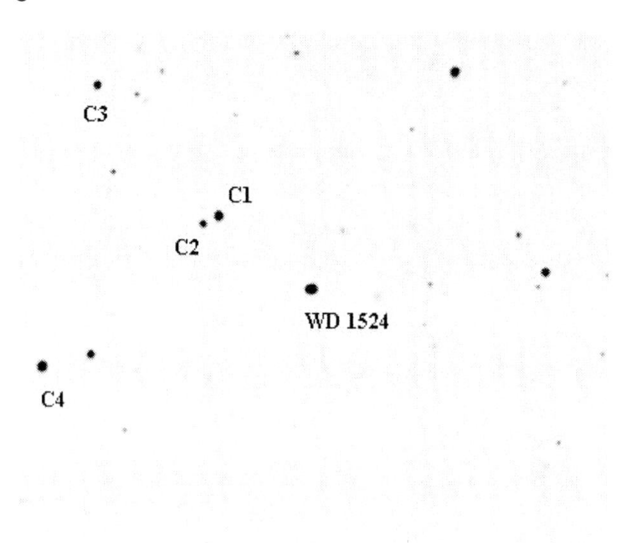

Figure 2: CCD Camera image of the observed field around WD 1524.

The field contains WD J1524-0030 (J2000: 15 24 03.25 -00 30 22.9) and four stars, marked as comparison stars C1, C2, C3 and C4. The exposure time was chosen to be 50 secs in R, V and B filters. Apart from the source observations, a large number of bias frames were taken before and after the source observations. Evening twilight sky flats in all the filters used in the observations were also taken. Standard procedures in the IRAF software, such as bias subtraction, flat-fielding and cosmic-rays correction, were followed for data reduction and analysis. On the clean images (bias subtracted, flat fielded), differential photometry was performed using C1 and C4 comparison stars along with source WD J1524-0030. Figure 3 shows the temporal variation of the relative magnitudes of WD J1524 with respect to these two comparison stars C1 and C4. The two differential light curves in Figure 3 are marked as (WD - C1) and (WD - C4), respectively. The X-axis shows observation number in the sequence images were taken as a function of time.

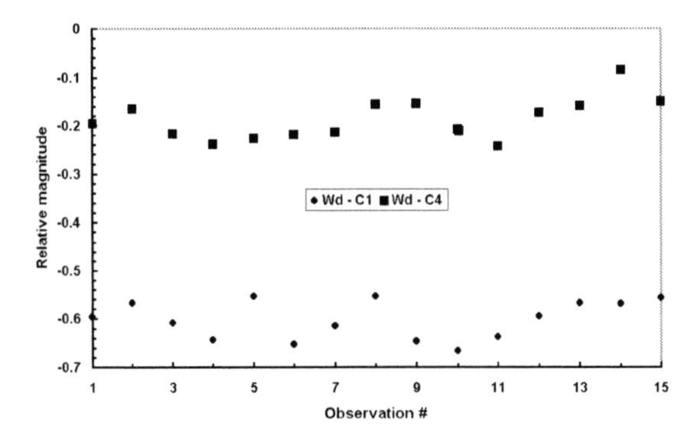

Figure 3: Temporal variations in the observed relative magnitude of WD 1524 with respect to comparison stars C1 and C4.

This preliminary photometric data collected for this source was processed using standard correlation and FFT software built in the IDL package. The auto-correlograms for the two series are shown in Figure 4. The Fourier transforms of the two data series of relative magnitudes are shown in Figure 5.

Both the spectra in Figure 5 show a peak at about the same frequency which is equivalent to a periodicity of 210 sec. As the data is of very short duration and sampling interval is 70 secs (including readout/overhead time), the resolution is low here and also the confidence limit will be poor. However, since the same periodicity is obtained from the relative magnitudes obtained by the two different comparison stars (C1 and C4), it gives some confidence in this estimate. Needless to say that longer monitoring with fast sampling will help attain better estimates of the periodicities in the source.

The present photometry results and the estimated periodicity are indicative that such observations and the procedure adopted can be used to provide vari-

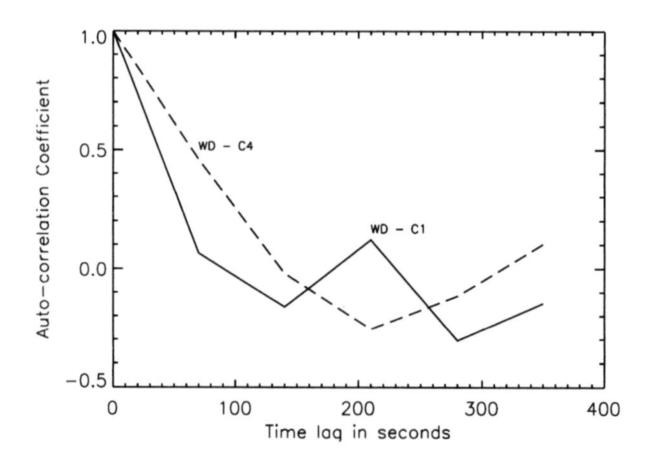

Figure 4: Auto-correlograms of the relative photometric data series.

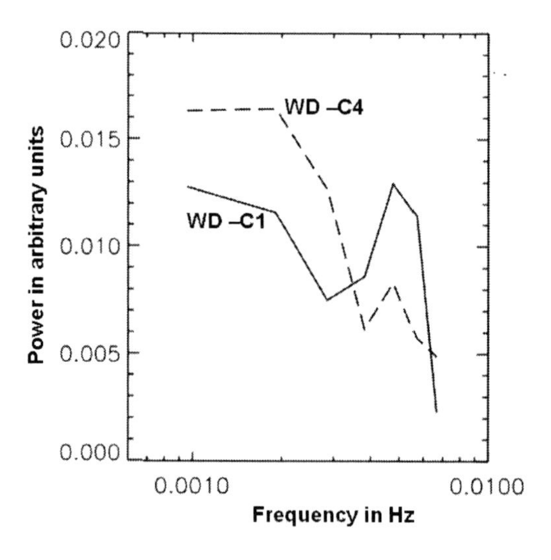

Figure 5: Power spectrum of relative photometric magnitudes.

ability information on WD J1524 and objects of this nature. In future we intend to participate in WET observing program to study pulsating white dwarfs, starting with November 2007 campaign on G38-29. The CCD instrument used in the present observations is not entirely suitable to study pulsating white dwarfs, because it has a substantial dead time (about 14 seconds) between exposures.

We are in the process of installing a fast CCD (EMCCD) for this type of observations which will enable us much faster sampling (with more than 20 frames per second) and more accurate photometry.

Conclusion

Though these observations on the WD1524 were in exploratory mode, we could see, from the optical photometry of very limited observation period, variations in the light-curve of the source. The Fourier transform analysis gives a period of about 210 seconds. We feel that present set up can be usefully employed for the WET Campaigns on white dwarfs in future. We also intend to use fast CCD (EMCCD) in future to overcome the large readout time problem.

Acknowledgments. Research at PRL is supported by the Department of Space, Gov. of India. One of us (HOV), thanks the organizers of DARC/WET Workshop 2007 to partially support his participation which provided much required interaction with active workers in the field.

References

Baliyan, K. S., Joshi, U. C., & Ganesh, S. 2007, presented at the INCURSI-2007, held at NPL, New Delhi, Feb. 21-24, 2007, AST-4:58

Deshpande, M. R. 1995, BASI, 23, 13

Mukadam, S. A., Montgomery, M. H., Winget, D. E., et al. 2006, ApJ, 640, 956

Winget, D. E. 1998, JPCM, 10, 11247

Comm. in Asteroseismology
Vol. 154, 2008

Introducing SPA, "The Stellar Photometry Assistant"

J. Dalessio[1,2,3], J. L. Provencal[1,2] & A. Kanaan[4]

[1] Delaware Asteroseismic Research Center, Mt. Cuba Astronomical Observatory, Greenville, DE 19807
[2] Department of Physics and Astronomy, The University of Delaware, 104 The Green, 217 Sharp Laboratory, Newark, DE, 19716
[3] Department of Astronomy, The Pennsylvania State University, 525 Davey Lab, University Park, PA, 16802
[4] Departamento de Física, Universidade Federal de Santa Catarina, CP 476, 88040-900, Florianópolis, SC, Brazil

Abstract

The Stellar Photometry Assistant, SPA, is a stand alone software package for time-series photometry reduction and analysis slated for an initial test release in spring 2008. The goal of SPA is to be simple, powerful, and intuitive. SPA was born out of complications in studying the pulsating DB white dwarf EC20058-5234 (QU Tel) due to the proximity of its nearby companions. SPA also addresses the Whole Earth Telescope's (WET) demand for large scale rapid data reduction from multiple sites. SPA is being developed in Matlab by the Delaware Asteroseismologic Research Center (DARC) in collaboration with the University of Delaware and the Mount Cuba Astronomical Observatory. The need for SPA is addressed, and key features of the program are listed and discussed.

The Need for SPA

QU Tel is a prime candidate for obtaining a $\dot{P}$ due to its rapid cooling rate and stable modes (Koen et. al. 1995). However, there are two nearby companions, one of which lies well within efficient aperture sizes. Due to the required precision to obtain a $\dot{P}$, combined with the general goal of reducing noise, more advanced data reduction techniques such as those described in Alard & Lupton (1998) and Stetson (1987) may be more suitable than traditional aperture photometry. IRAF is capable of such methods, but the complexity of these methods using IRAF can be challenging to inexperienced users. While SPA is currently only capable of aperture photometry, the modular structure of SPA will allow future implementations of more powerful methods. Meanwhile, the WET obtains large sets of photometric data for several objects semi-annually. The large quantity of data, and the variety of observing sites can make data reduction challenging and tedious. During a typical WET observation, multiple users participate in data reduction around the clock. As data is analysed, time critical decisions are made selecting targets and coordinating observations. This makes rapid data reduction a necessity. A tool like SPA would ideally lessen the manpower required at headquarters for a WET observation. This would ease operation of the WET financially.

Program Features

An Intuitive Interface

SPA features a graphical user interface with tool tips and help menus. SPA has been designed with the hope that users new to time-series photometry will find it easy to use while advanced users will not suffer from a lack power. Users first build or load parameter files (mentioned below). They then select an entire directory contents or specific files to be processed. If calibration files are selected they are applied in a user specified manner. The structure of the interface is, however, constantly under revision.

Smart Star Finding

SPA will only require user entry of star pixel positions from one image per filter. SPA will build a "star orientation file" containing these positions and calculate relative object intensities. SPA can then automatically acquire/reacquire the objects for any full or partial image of this field when that "star orientation file" is selected. This database can be used for other telescopes, even when the field of view is substantially different (as long as the field of view is specified). This allows large jumps in star positions to be easily handled with no interruption as well as lessening the amount of user input required for reduction of multiple observations of the same field (as in the case of the WET). This method will allow for both translational and rotational variance from the previously known object pixel coordinates. Currently SPA searches for a local maximum near the last known position of the object and centers with a simple model PSF. More advanced methods are still in development and automatic star finding may not be included in the initial test release.

FITS Header Flexibility

Fits headers vary site to site and this presents difficulties reducing data for large scale collaborations such as the WET. SPA allows users to enter location information into a database. This includes information about header formatting. When processing a data set SPA can either detect the location from the fits headers automatically or the location can be selected from the database. This eliminates the need for managing and editing fits headers.

Portability

SPA will be available on a variety of platforms. The goal will be to have a free standing program with a simple single file distribution (including the Matlab Runtime Component). Functionality will be identical on all platforms. Public availability of the source code is under discussion. While contribution from the community will be valuable, it may be advantageous to restrict source distribution in order to be able to simplify supporting the software.

Standardized Output

SPA also builds a "SPA output file" (*.SPA) including the photometry parameters used, calibration information, various data from the light curve (raw counts, sky counts, etc.), processing logs, along with any analysis performed on the data. These files are exchangeable and can be loaded by another SPA user. SPA can export data from these files into an ASCII table. SPA also allows quick visualization of data contained in these SPA files.

Analysis Tools

Currently, the only analysis tool built into SPA, besides plotting, is the discreet Fourier transform (DFT). However, additional tools may be added before the initial release, such as pre-whitening.

Testing the Effectiveness of SPA

IRAF scripts developed by Kanaan et. al. (2002) are being used to test the reliability and effectiveness of SPA. Simple aperture photometry produces similar light curves. Differences in the light curves are marginal and are likely due to variations in the methods chosen to round pixels or center an object. At first glance SPA seems substantially better at tracking stars in crowded fields. A more rigorous comparison is planned prior to the initial release.

Acknowledgments. Special thanks to the WET team and the Mount Cuba Astronomical Observatory for their support.

References

Alard, C., & Lupton, R. H. 1998, ApJ, 503, 325

Kanaan, A., Kepler, S. O., & Winget, D. E. 2002, A&A, 389, 896

Koen, C., O'Donoghuee, D., Stobie, R. S., et al. 1995, MNRAS, 277, 913

Stetson, P. B. 1987, PASP, 99, 191

J. Xiaojun informs the WET about the BAO telescopes.